Open access edition supported by the National Endowment for the Humanities / Andrew W. Mellon Foundation Humanities Open Book Program.

© 2019 Johns Hopkins University Press
Published 2019

Johns Hopkins University Press
2715 North Charles Street
Baltimore, Maryland 21218-4363
www.press.jhu.edu

The text of this book is licensed under a Creative Commons Attribution-NonCommercial-NoDerivatives 4.0 International License: https://creativecommons.org/licenses/by-nc-nd/4.0/.
CC BY-NC-ND

ISBN-13: 978-1-4214-3592-3 (open access)
ISBN-10: 1-4214-3592-6 (open access)

ISBN-13: 978-1-4214-3590-9 (pbk. : alk. paper)
ISBN-10: 1-4214-3590-X (pbk. : alk. paper)

ISBN-13: 978-1-4214-3591-6 (electronic)
ISBN-10: 1-4214-3591-8 (electronic)

This page supersedes the copyright page included in the original publication of this work.

Supplying the Nuclear Arsenal

The Johns Hopkins University Press | Baltimore and London

Rodney P. Carlisle

with Joan M. Zenzen

SUPPLYING THE NUCLEAR ARSENAL
American Production Reactors, 1942–1992

© 1996 The Johns Hopkins University Press
All rights reserved. Published 1996
Printed in the United States of America on acid-free paper

05 04 03 02 01 00 99 98 97 96   5 4 3 2 1

The Johns Hopkins University Press
2715 North Charles Street
Baltimore, Maryland 21218-4319
The Johns Hopkins Press Ltd., London

Library of Congress Cataloging-in-Publication Data will be found at the end of this book.

A catalog record for this book is available from the British Library.

ISBN 0-8018-5207-2

CONTENTS

	List of Illustrations	vii
	List of Tables and Figures	ix
	Preface	xi
	List of Abbreviations and Acronyms	xiii
	Introduction	1
1	Inventing Atomic Piles	8
2	Building Reactors at Hanford	26
3	Contracting Atoms	46
4	Flexible Design at Savannah River	67
5	The Arms Race Arsenal	92
6	Designing a Reactor for Peace and War	107
7	Surviving Détente	131
8	Lobbying for Nuclear Pork	164
9	Managing Nuclear Options	195
	Conclusion: Supplying the Cold War Arsenal	219
	Appendix: Production Reactor Families	225
	Notes	229
	Bibliographic Essay	263
	Index	269

ILLUSTRATIONS

Enrico Fermi explaining a formula to fellow scientists	15
The Hanford B reactor	40
The fuel-element face of B reactor	41
Gen. Leslie R. Groves and David E. Lilienthal	49
The Savannah River R reactor	82
Schematic of piping for heavy-water reactors at Savannah River	83
Construction of vessels for a Savannah River reactor	89
The hybrid Hanford N reactor	112
The charging face of N reactor	125
The N reactor control room	127
Map of the Hanford site	155

TABLES AND FIGURES

Tables

1	Reactor scrams in early operations at Hanford	42
2	Postwar production reactor completion schedules	78
3	Power levels at Hanford reactors, 1958	104
4	Summary of power and production reactors, 1951–1963, by type	114
5	Power level upgrades of production reactors	147
6	Reactor closings during the plutonium glut period, 1964–1971	152
7	Nuclear arms control treaties, 1963–1979	166
8	Relative seniority of U.S. senators from reactor-site states	169
A.1	U.S. production reactor families	226
A.2	Auxiliary test and experimental reactors at Hanford and Savannah River	227

Figures

1	Time line of production reactor construction schedules, 1950–1955	98
2	Reactors for war and peace, 1963–1976	153

PREFACE

This work derives from a manuscript originally completed under contract between the Department of Energy's Office of New Production Reactors (ONPR) and History Associates Incorporated (HAI), which received funding over three years to prepare a history of production reactors. During this period, HAI staff gathered copies of pertinent documents from the National Archives, from the Department of Energy Archives, and from record groups at Richland, Washington, and Wilmington, Delaware. In addition, HAI researchers gathered documents from separate offices within ONPR. We met several times with the director of ONPR, Dr. Dominic Monetta, and his administrative assistant, Michael Shapiro, together with management consultants Dr. D. Scott Sink and Dr. Harold Kurstedt from Virginia Polytechnic Institute and State University. These meetings were invaluable for providing insight into current operations, management approaches, and the day-to-day decision-making process and for helping to develop a more thorough understanding of the evolution of nuclear engineering. Further meetings with Monetta's successor, Tom Hendrickson, shed light on the final days of the New Production Reactor effort. ONPR ceased operations early in 1993, and we completed and submitted the completed preliminary manuscript history to the Department of Energy's History Division.

Both ONPR and the History Division allowed us to shape and define the history and to interpret the facts as we saw them, providing us with suggestions for clarification. The only constraints were the length of the manuscript and the amount of time and resources we could expend.

As in any work over a period of time, the authors owe a series of intellectual debts to a number of people who assisted in the research and who provided readings of part or all of the manuscript. In our case, that debt was compounded by the fact that we worked through History Associates

Incorporated, which brings a team approach to its tasks, and through the Department of Energy. Within HAI we had able research assistance from Laurie Kehl, Kathryn Norseth, Teresa Lucas, Michelle Hanson, Adam Hornbuckle, Jim Gilchrist, Jonathan Koenig, Greg Wright, and Eric Golightly. James Lide not only provided research but prepared an early draft of some materials incorporated in Chapter 8. Readings and suggestions from Ruth Dudgeon, Ruth Harris, Brian Martin, and Richard Hewlett led to direct improvements; Gail Mathews and Darlene Wilt assisted in production of the preliminary manuscript. Our research at a number of facilities was assisted by the depth of knowledge of the archivists, including Marjorie Ciarlanti at the National Archives in Washington, D.C.; Dr. Michael Nash, Marjorie McMinch, and Lynn Catenese at the Hagley Museum; Flo Ungefug at the Records Holding Area at the Hanford Operations Office in Richland, Washington; Terri Traub at the Hanford Public Reading Room; and Dr. Roger Anders at the holdings of the Department of Energy in Germantown, Maryland. At the Department of Energy, readings and comments by Drs. Frank Cooling and Roger Anders sharpened and corrected a number of points in the manuscript. Jane Register and Rich Goorevich facilitated our access to personnel and files at ONPR. We extend our thanks to all of these folk.

In the Francis Parkman tradition, we also had the unique opportunity to tour three Hanford production reactors and gain insight from seeing the control rooms, reactor faces, and other equipment. We thank Mike Berriochoa for arranging the tour, Don Lewis for sharing his experiences in operating B reactor during World War II, and Herb Debban for leading the tour, which brought home the differences between the earlier Hanford reactors and N reactor.

As to our co-authorship, Dr. Rodney Carlisle took the lead in writing on this project, and Ms. Joan Zenzen supervised the various research assistants, organized the voluminous files collected, and wrote two chapters of the preliminary manuscript as well as editing that version. Carlisle reviewed and revised the whole manuscript for the publication with Johns Hopkins University Press.

The patience of our spouses, Loretta and Stuart, reached heroic proportions.

ABBREVIATIONS AND ACRONYMS

A&E	architectural and engineering (firm)
ACRS	Advisory Committee on Reactor Safeguards
AEC	Atomic Energy Commission
BORAX	BOiling water ReActor—eXperimental
CEGA	Combustion Engineering-General Atomics
CP	Chicago Pile
CSSAP	Concept and Site Selection Advisory Panel
DOE	U.S. Department of Energy
DP	Defense Programs (office within DOE)
DPR	dual purpose reactor
DPW	Du Pont—Wilmington
EIS	environmental impact statement
ERAB	Energy Research Advisory Board
ERDA	Energy Research and Development Administration
FFTF	Fast-Flux Test Facility
FFTR	Fast-Flux Test Reactor (also: Fast-Fuel Test Reactor)
GAC	General Advisory Committee
GE	General Electric
GOCO	government-owned, contractor-operated (facility)
HEU	highly enriched uranium
HTGR	high-temperature gas-cooled reactor
HW	heavy water (deuterium)
HWCTR	Heavy-Water Components Test Reactor
HWR	heavy-water (-cooled and -moderated) reactor
I&E	Internally and externally cooled (slugs)
INEL	Idaho National Experimental Laboratory
JCAE	Joint Committee on Atomic Energy
kW	kilowatt
LMFBR	liquid-metal fast breeder reactor
LTHWR	low-temperature heavy-water (cooled and -moderated) reactor
LWR	light-water (-cooled and -moderated) reactor
MAD	mutual assured destruction

MED	Manhattan Engineer District
Met Lab	Metallurgical Laboratory
MHTGR	modular high-temperature gas-cooled reactor
MIRV	multiple independently targeted reentry vehicle
MLC	Military Liaison Committee
MW	megawatt
MW(e)	megawatt (electrical)
MW(t)	megawatt (thermal)
NAE	National Academy of Engineering
NARA	National Archives and Records Administration
NAS	National Academy of Sciences
NASA	National Aeronautics and Space Administration
NEPA	National Environmental Policy Act
NIF	Naval Industrial Funding
NPR	new production reactor
NRC	Nuclear Regulatory Commisison
ONPR	Office of New Production Reactors
OSRD	Office of Scientific Research and Development
P-9	"Product 9" (code name for heavy water)
PRA	probabilistic risk assessment (or analysis)
Pu	plutonium
PWR	pressurized-water reactor
RAMI	repairability, accessibility, maintainability, and inspectability
RBMK	(Acronym from Russian) Graphite-moderated steam-generating reactor of the type used at Chernobyl
RNR	replacement N reactor
RI	reactor incident
ROD	Record of Decision
RSC	Reactor Safeguards Committee
SALT	Strategic Arms Limitation Treaty
SET	site evaluation team
SNAP	space nuclear auxiliary power
SRS	Savannah River Site
START	Strategic Arms Reduction Treaty
TRIDEC	Tri-Cities Industrial Development Economic Council
TVA	Tennessee Valley Authority
U	uranium
WNP	Washington Nuclear Power
WPA	Works Progress Administration (Works Projects Administration)
WPPSS	Washington Public Power Supply System
ZEPHR	zero-electric-power heavy-water reactor

Supplying the Nuclear Arsenal

Introduction

For the 50 years between 1942 and 1992, the United States centered its military and foreign policy on nuclear weapons. These weapons ended World War II, contributed to the origins of the Cold War between the Soviet Union and the Western Allies, defined the arms race and the doctrine of deterrence, and were the ultimate bargaining chips in a diplomatic dance to achieve arms control and détente between the United States and the Soviet Union. As the Cold War ended, these weapons continued to threaten the uneasy and troubled peace that ensued, leaving a legacy of risk.

The American scientists and European-emigré scientists living in the United States who built the first bombs knew well that Germany was fully as capable of developing an atomic bomb as the United States. As it turned out, Germany did not pursue the development of atomic weapons with the needed resources. Had the Nazi leadership learned of the magnitude of the effort the United States devoted to making the bomb, however, all of that could have changed. So the cloak of secrecy dropped over the facilities at Chicago, Illinois; at Los Alamos, New Mexico; at Oak Ridge, Tennessee; at Hanford, Washington; and at dozens of smaller laboratories and plants scattered around the nation.

Nuclear spies from the Soviet Union succeeded in penetrating this screen, yet the physical principle of the bomb was elementary, and espionage only supplemented the work of Soviet physicists who sought to harness nuclear energy. From the first, physicists predicted that the American monopoly on the atom bomb would be temporary and that they had unleashed a long-term horror upon the world. They were right. By the 1990s, the logic and imperatives of physics spawned nuclear weapons in at least ten nations with others working quietly to join the "atomic club."

The specter of nuclear holocaust still haunted the world. Where the Soviet Union had represented one nuclear power, four of the states that emerged from the Union's breakup—Russia, Byelarus, Ukraine, and Kazakhstan—were in a position to export, legally or illegally, the weapons and know-how. Rogue nations like North Korea, Iraq, and Libya worked to acquire small but deadly arsenals.

Ironically, the nuclear powers' secrecy about the detailed mechanics of building such weapons hampered open discussion of the issues over the decades. In the United States, members of Congress, journalists, administrators, and the reading public remained generally uninformed about even the basic procedures by which the weapons were created. Information long declassified and published remained out of the public eye, often sheltered by the culture of secrecy. The public increasingly came to distrust experts who made decisions behind the closed doors of technical knowledge and classified information.

In 1945, the government had released many of the essentials in the Smyth Report. Americans developed two basic designs of atomic bombs during World War II. One used a rare and difficult-to-extract isotope of uranium, U-235. The other design, the one tested in the first nuclear explosion over the desert at Alamogordo, New Mexico, and used for the weapon detonated over Nagasaki, Japan, depended upon a new man-made element, plutonium, with an atomic weight of 239. Plutonium-239 was made in machines modeled on one invented in 1942, an atomic pile or nuclear reactor.

All later nuclear reactors used for the generation of electrical power, for propelling submarines and ships, and for research could trace their ancestry back to the pile first operated in 1942 in Chicago and invented by the nuclear physicist Enrico Fermi. For the next 50 years, all plutonium used in the United States to supply the nuclear arsenal was manufactured in 14 "production reactors," industrial-scale machines that upgraded or improved upon the Fermi pile. Thus, production reactors have a longer history by about two decades than do the better-known and better-documented power reactors.

The 14 piles, or reactors, that were designed to produce plutonium were at the very heart of the nationwide industrial enterprise built to make atomic weapons. Despite their crucial importance to America's nuclear effort, and despite the fact that any proliferation nation, such as North Korea, could build similar machines or convert other reactors to such a function, the informed public was barely aware of their existence or their mechanics.

This book tells the story of the 14 machines at the essential center of the American nuclear weapons complex. How were those 14 machines designed, built, and managed? How were they modified to keep up with the demands of the nuclear arsenal? As they aged, what plans were made to replace them? What politics and policy drove the machines, and how did they, in turn, drive politics? At a broader level, how did the modern American political structure deal with decisions regarding such massive technological efforts?

In order to study design, construction, management, and policy, and to understand how specific technological choices were made or delayed, we examined the voluminous records of the Manhattan Engineer District, the Atomic Energy Commission, the Energy Research and Development Administration, and the Department of Energy. That sequence of governmental institutions developed its own values, standards, vocabulary, symbols, and myths. The organizational structure of the weapons complex created a culture unique in America, a culture itself subject to attack by the 1980s.

The sources of that culture explained some of its characteristics. The industrial, military, and scientific communities of the United States had to come together in World War II to build the piles, to produce the uranium and plutonium, and to make the bombs. The cooperation was contentious and awkward. Each group had its own values, sometimes at odds with the others. General Leslie R. Groves and other officers from the Army Corps of Engineers directed the Manhattan Engineer District, which built the reactors and the bombs. Physicists from Columbia University, the University of Chicago, the University of California, and other institutions left their laboratories to do design and actual construction work. Du Pont provided chemical engineers and business administrators to plan and manage the industrial-scale reactors that would become plutonium-manufacturing facilities. This cumbersome assemblage of men and women from different backgrounds was necessary to make the weapons quickly, under tight wartime schedules, and to prevent any knowledge of the program from reaching Germany.

The unique nuclear-weapons culture emerged out of the values and management styles brought by the scientists, businessmen, and army officers. In the postwar era, that evolving culture continued to define and shape how the production reactors of the Cold War would be managed. The cloak of secrecy was wrapped closer as news of Soviet advances and Soviet espionage fed American fears. Those fears led to the stripping of a security clearance from J. Robert Oppenheimer, the scientist who had directed the bomb-design effort at Los Alamos. In that crisis, the nation had a public glimpse of one aspect of the internal cultural conflict which continued to seethe within the nuclear establishment. The internal struggle found echoes in national ideological and political alignments.

The career incompatibility of business executives and army officers also had important, but somewhat less spectacular long-range political consequences. In order to meet the salary demands and expectations of the engineers and corporate managers, the Army set up a system of working through long-term contracts with the private sector. The nuclear establishment soon became a lucrative source of employment for tens of thousands of administrators, managers, supervisors, scientists, engineers, technicians, and support staff. Scattered across the nation in geographically isolated communities, these groups eventually became articulate politically, with active spokesmen and defenders in Congress and the Senate.

When in 1949 the Soviet Union detonated its first nuclear weapon, the American response included an accelerated search for ways to make even more powerful weapons. A key ingredient in boosted atomic bombs and in the hydrogen bomb was an isotope of hydrogen which could be best manufactured in nuclear reactors: tritium. Tritium decayed by a process of radioactive emission. Its half-life was relatively short, meaning that the stockpile of produced tritium had to be constantly replenished if the American nuclear arsenal were to be maintained. Thus, by the 1950s, to keep a supply of workable and effective weapons in the Cold War arsenal, planners found it essential to expand and upgrade the production reactor family at the heart of the nuclear arsenal. Over the period 1949–64, in several waves of construction, the United States added 11 more reactors.

As soon as all 14 reactors were operating, weapons planners recognized that the earliest reactors, built hastily in the war years, had become increasingly risky. Furthermore, the nation's supply of plutonium was more than adequate for current and projected weapons, although a continued supply of tritium had to be assured. During the 1960s, the AEC ordered the closure, one by one, of most of the old reactors. Each closure resulted in laying off hundreds of employees. Contractors, local merchants, politicians, union leaders, and self-appointed advocates of those dismissed employees fought to keep the reactors operating as long as feasible. The pain of closure, rather than the benefit or risk of opening, was one of the first issues to force the decisions of production-reactor policy out from behind the walls of secrecy into the open world of media and congressional debate.

By the early 1970s, the cousins of the production reactors—nuclear reactors built for the peacetime purpose of generating electrical power—encountered organized political resistance, at first on a local and regional basis, based on legitimate fears of risk and on suspicion of the motives of managers. By the mid-1970s a more broadly based and popular antinuclear movement drew support; some decisions to build power reactors encountered opposition and even angry protest. Accidents, media interest, and growing public information about the nature of reactor risks meant that siting and technical choices for both commercial power reactors and production reactors could no longer be made outside the public's view.

The issue of where to site a commercial reactor often ran into the "not-in-my-back-yard," or "NIMBY," reaction, depending upon the relations between the utility and its neighbors and consumers. Some commercial reactors stirred very little opposition; with others, advocates and opponents struggled over issues of safety, risk, employment, economy, and arms and disarmament. Pollution of waterways, threats of airborne radioactive emissions, the impact of waste handling and ordinary construction activity upon endangered species—such issues affected the choice of site and design for some of the commercial reactors and drew attention to similar concerns for production reactors. For some, power reactors became surrogates for the threat of nuclear destruction implicit in the Cold War; opposition to a power reactor could also express one's opposition to that more horrible nuclear risk.

As détente held out the hope of arms control and as the 14 production reactors aged and were gradually shut down, politicians and nuclear planners debated whether and how to replace them with one or more new production reactors. That debate pitted the emergent interest groups against each other in a protracted and often bitter conflict. Communities and states desiring the federal expenditure aligned with industrial concerns in arguing for their share of "nuclear pork." Opponents of the arms race explicitly and openly aligned with advocates of protecting the environment against the risks of nuclear emissions and nuclear accident in fighting the plans. Some of the open political debates were flavored with the language of the technopolitical debates among experts.

American nuclear policymakers faced a complex set of interrelated decisions. How should they address the anticipated shortfall of tritium which would result if no reactors replaced the rapidly decaying isotope essential to the modern weapons? Should the nation undertake the enormous expense necessary to build another production reactor? If so, should planners follow an innovative or a well-tried design? Where to site the next generation of production reactors, which technology to plan, and when to build became matters of congressional debate. Articulate advocates argued for one or another site or concept; consortiums of corporations devoted to different reactor types fought for the prize; competing representatives and senators spoke out with force for constituents whose fortunes and neighborhoods would be heavily affected. For nearly a decade, it appeared that the process of democratic debate would prevent action and that tritium decay would begin to shrink the nuclear arsenal. The prospect encouraged antinuclear advocates and represented a threat to security for strategic thinkers.

By the mid-1980s, Congress seemed deadlocked over the needed decisions. In 1942, General Groves had selected the wartime sites on the basis of nearby resources, risk to populations, available labor supply, and other criteria. By the 1980s, however, each issue, such as resource, risk, or potential payroll, when opened to public or political discussion, brought a scramble for contracts and employment; a public airing of issues of war and peace, earth and man's place in the natural environment; and arcane debates over probabilistic analyses of risk and alternate engineering styles and designs. Thus, by the 1980s, nuclear politics had three cross-cutting alignments: traditional pork-barrel advocates defending the interests of the thousands of employees of existing and potential contractors, public interest groups fighting over issues of environment and disarmament, and the technopolitical advocates of one or another design.

To build even one new production reactor would require a project as complex, as expensive, and as demanding of new management approaches as Polaris, which produced submarine-launched ballistic missiles, or Apollo, which put Americans on the surface of the moon. To some, the lengthy and fruitless debate over the next generation of production reactors demonstrated that the American polity had become too fragmented to take on an engineering challenge of such a scale. Although General Groves had chosen site, concept, machinery, and contractor, all in a few months, changed conditions by the 1980s seemed to immobilize those charged with making analogous choices. Technical progress appeared stymied by political confusion and the turgid working of democracy.

In an attempt to sort through the technical decisions in an objective, nonpolitical way, the Department of Energy mounted a concerted planning effort. A specially assembled office gathered all the information and sifted through it under rigorous objective standards designed to protect the public interest. The Department of Energy made an explicit effort to break with some of the cultural imperatives of the weapons complex and even adopted the language of cultural analysis in the attempt.

The mere fact that the effort began to absorb several hundred million dollars a year of design and management effort may have carried weight in arms control negotiations with the Soviet Union. Such a show of planning to build a new reactor indicated to the Soviets that the United States had a serious willingness to continue the expensive arms race. Suffering internal disarray, the USSR suddenly began to negotiate a genuine mutual downgrading of the massive weapons arsenals.

To many observers through the Cold War period, it seemed that technology itself drove the arms race. The competition between the Soviet Union and the United States to increase the number, effectiveness, and deliverability of nuclear weapons appeared to derive from the nature of technical advance itself and served as a dreaded reminder of how machines had come to dominate human affairs. The arms race seemed inevitable, the technologies of the weapons drove requirements for supply of material, the need for the material drove upgraded reactor designs, and the new machines produced ever-increasing numbers of weapons, stimulating a similar response from the Soviets.

And yet, as the planning for the next generation of production reactors went forward, outside policy events put the reactors themselves on the endangered species list. The end of the Cold War, the dissolution of the Soviet Union, and the impact of those political events on weapons policy showed that the machines and weapons were, after all, only tools of policy, not the drivers of that policy. Under new arms control treaties, the disarmament imposed by diplomats would move at a faster rate than the pending disarmament imposed by the half-life of tritium. The new reactors would not be needed unless or until a new nuclear arms race emerged from the ashes of the Cold War. Early in the 1990s, the effort to replace the old production reactors paused.

This work is written in that pause, in hopes that we can better understand the interplay of weapons, the machines that make them, and the human cultures that grow up around those machines in retrospect.

1 • Inventing Atomic Piles

The design and construction of the first production reactors to make plutonium for atomic bombs in World War II is a story remarkable not least of all because of the speed with which the task was completed. From conception to production, the massive project took less than two years; within another year the weapons had been designed, tested, and delivered. It was early in 1942 that physicists concluded that plutonium might be produced in sufficient quantities for nuclear weapons in industrial-scale nuclear reactors and began to consider the various designs of such devices. In December 1942, Enrico Fermi achieved the first self-sustaining nuclear chain reaction in the reactor later known as CP-1. Within two months a contractor had been chosen and a site had been selected at Hanford, Washington, for the production reactors; construction began by mid-1943. The contractor built three reactors and placed them all in operation by February 1945. Hanford-produced plutonium was in the first nuclear explosion at Trinity, 16 July 1945, and in the weapon exploded over Nagasaki, 9 August 1945.

Over the same years, the United States launched 50 million tons of merchant shipping, built over 300,000 aircraft, and successfully developed synthetic rubber, radar, proximity fuses, the bazooka, barrage rockets, the "wonder drugs" sulfa and penicillin, and new pesticides. The speed and magnitude of American construction of nuclear weapons and of the production facilities behind them was part of that remarkable and concerted effort to enlist science, engineering, industry, and labor in the war effort. As memoirs and published diaries recounted the wartime achievements, a growing body of literature focused on the personalities, the force of strong leadership, and the creation of new management structures. Applied science won the war, but successful leadership converted that science to ships, airplanes, rockets, and bombs.[1]

Looking closely at the production reactor role in wartime mobilization from the perspective of the 1990s prompts several new questions. In the light of later management concerns about nuclear reactors, both for weapons material production and for power generation, it is natural to ask how the wartime generation dealt so quickly with issues which in later decades would take far longer to resolve. By the 1980s a set of intricate, often time-consuming procedures had evolved around each of the issues. Fifty years of experience with the technology has yielded progress in design techniques, in materials sciences, in fundamental knowledge of nuclear processes, and in management structure. Even so, what took two years in the 1940s would take at least ten years in the 1980s and 1990s.

At one level, the reason is simple. War compresses research and development time. In 1942, American progress was driven by the fear that Hitler's Germany would develop the nuclear weapon first. But no matter how pressing the need, some of the issues could not be ignored; steps and stages in the complex process had to be taken, not eliminated. Even in the urgency of a wartime arms race, General Leslie Groves, who commanded the Manhattan Engineer District (MED) from 1942 through 1946, had to take the decision we now call site selection. In one way or another, Groves drove the project through the steps a later generation would categorize as conceptual design, design review, contractor selection, safety analysis, and risk assessment. Groves and his associates evaluated and mitigated the environmental impact of reactors. None of these stages were described in such bureaucratic language in the 1940s, yet each step was accomplished in weeks or months in 1943 and 1944.

The Culture of the Manhattan Engineer District

An understanding of how the culture of the weapons complex developed gives a needed perspective on the technology itself. An examination of the background makes more clear how particular groups, acting under particular cultural values, made specific choices about the design of production reactors.[2] Over and over, specific arrangements which Groves and his team worked out provided the organizational seeds from which a whole forest of practices and procedures later grew, so that what had been established during wartime eventually became the unique organizational culture of the nuclear weapons complex. For example, a number of laboratories, administered by universities and usually headed by a prominent and established physicist, were created across the country. The industrial parts of the complex, also scattered around the nation, were administered under contract by major manufacturing corporations under the direction of corporate-employed engineer-administrators. Each of the locations was subject to military rules of security, with armed guards checking identification and traffic; information was shared on a need-to-know basis. In later decades this organizational structure and culture would resist, and sometimes grudgingly adapt to, the changed political requirements of civilian administration, détente, participatory government, and popular environmental concerns.

Physics had been enlisted in the war effort, and then in peacetime policymaking, became part of a broader development that took experts from academia into the heart of military and governmental action. In the area of nuclear policy, especially, academics became administrators. As time went on, however, the authority of nuclear experts became increasingly questioned and eroded over issues of safety and the environment.[3]

The origins of many of the specific institutional aspects of the nuclear establishment, and the origins of the culture that those institutions carried, can be seen in the first arrangements of MED. The way of doing business which Groves established was efficient partly because it did not require outside participation. In the interest of keeping secret from the enemy the very fact that the United States had decided to pursue the possibility of atomic weapons, all of the structures, both physical and organizational, were created without publicity and without congressional knowledge.

The culture survived and evolved under the successor agencies that later managed the nuclear establishment: the Atomic Energy Commission (AEC) (1946–74), the Energy Research and Development Administration (ERDA) (1975–77), and finally, the Department of Energy (DOE). Under the aegis of all these agencies, the weapons complex continued the practice of the 1940s in a series of compromises between academics and private industry, between scientists and engineers, between civilians and military men. But whereas before 1945, secrecy was so thorough that no compromises were required with Congress or with the general public, those adjustments would have to be made by the successor agencies.[4]

One of the most striking characteristics of the nuclear establishment since World War II has been the peacetime use of industrial contractors to operate large government-owned laboratories and industrial facilities. The government-owned, contractor-operated (GOCO) establishments at Hanford,

Washington, and at Oak Ridge, Tennessee, were later supplemented by others at Albuquerque, New Mexico; Amarillo, Texas; Portsmouth, Ohio; Paducah, Kentucky; and elsewhere. The direct MED-operated laboratory at Los Alamos was operated on contract by the University of California; Chicago's Metallurgical Laboratory, after some evolution, emerged as Argonne National Laboratory, under the administration of the University of Chicago. Other universities, either singly, or in consortiums, operated later laboratories such as Brookhaven and Lawrence Livermore.

Before MED created new patterns, peacetime scientific and technical research and industrial enterprises for the military had usually been operated under older, traditional arrangements: government-owned and -operated facilities such as the Navy's Powder Factory at Indian Head, Maryland, or contractor-owned and -operated facilities, like Du Pont's Wilmington laboratories.[5] Scientific projects had been funded under both grants and contracts, but the continued operation of federally owned laboratories by universities was a new departure. Groves's system of operating the industrial side through a major corporation and running the scientific side through a major university became permanent organizational features of the new nuclear establishment. In business terms, the operating contract through which the nation's nuclear laboratories and factories were managed represented an institutional innovation. The GOCO had precedents in government-financed industrial facilities such as steel armor plants for the Navy and gun-forging plants for the Army. However, the nuclear GOCO that characterized the weapons complex over the next 50 years was born in the 1942 contracts arranged to build and operate the industrial-scale facilities needed to produce the first bombs.

Another of Groves's arrangements which lived on to characterize the postwar nuclear establishment was his introduction of academic physicists, who were used to the open world of scientific conferences and publication in journals, into the closed world of secrecy in which the activities of various specialists were kept isolated from each other and their work "compartmentalized." American chemical engineers were familiar with the constraints of secrecy, in both proprietary industrial settings and government laboratories, before World War II. But physicists, like philosophers and mathematicians, had thrived on open publication of their theoretical work. Before 1940, almost all theoretical physicists pursued their work in universities and colleges; they constituted a truly international fraternity of scientists, publishing and reading each other's material across national and language boundaries. The project to use nuclear physics to make weapons would bring down a curtain of secrecy. Surprisingly, it was some of the most outspoken and independent physicists, like Leo Szilard, who insisted on the need for secrecy, anticipating the significance of atomic research for weapons development.[6]

Groves "married" physics to engineering in other ways. At Los Alamos, scientists engineered the bomb. At Hanford, engineers enlarged the pile that physicist Fermi had built to demonstrate a principle, redesigning the concept into industrial-scale devices in Hanford's production reactors. Out of these beginnings, supplemented by later organizational developments in the field of naval reactors, a new profession eventually emerged, that of nuclear engineering.[7]

Cooperation between physicists and engineers, academics and industrial businessmen, and civilians and military officers was difficult to achieve. The wartime project brought together individuals with quite different mind-sets and with different kinds of cultural baggage. While each group contained a wide range of individuals with different values and preconceptions, members of each group tended to hold stereotypes of the other groups. Some of the corporate engineers from Du Pont thought the scientists impractical dreamers; a group of the Chicago scientists found the corporate engineers too hasty and lacking in academic grounding. Heated disagreements flared over management style and the correct assignment of responsibility among theory, research, design, engineering, training, and operation. Personality differences, prejudices, and bickering sometimes escalated disagreements into crises; on the other hand, a few diplomatic managers tried to keep everything running smoothly towards the goals.

If some of the stereotypes—the absent-minded professor, the pragmatic corporate official, the small-minded martinet of a military officer—sound hackneyed and familiar, it is because they were entrenched in the popular culture of the 1930s and 1940s. To varying degrees, nonacademics viewed professors as egotistical, impractical, and given to long-winded speculation. Civilians viewed military officers as unimaginative and lacking initiative. Some found the military's concern with security stifling to creativity. A few cantankerous scientists thought engineers incapable of understanding the theoretical issues involved and the need for experimentation, although practical physicists and well-trained engineers often had similar education, careers, and values. Academics frequently suspected corporate leaders of seeking monopoly and profit. The scientists, Groves noted in his memoir, "particularly those educated in Europe," distrusted corporations, and they "had the idea that all design and engineering for the project should be accomplished under their personal direction."[8]

The conflicts and tensions arose not only because men and women from these different backgrounds held prejudices about each other but also because they brought different ways of doing business from their disparate backgrounds. The organization that emerged reflected sometimes conflicting elements of the various operating styles, blending military secrecy, business methods, and academic research. Corporate leaders from Du Pont conducted business on company pay scales, using company organizational structures, and ensuring company seniority for employees. In particular, Du Pont brought its own style of development, combining flexible and multiple approaches to problems and the "freezing" of design of certain components while others proceeded, a procedure which had been proven through the mid-1930s in Du Pont's development of nylon.[9] Day-to-day administrative procedure drew from the various backgrounds of military, academic, and corporate cultures. Badging, fences, and guards followed Army guidelines. Fermi and his colleagues continued to work through loose, overlapping committees, much as they had in academic departments, with senior colleagues assisted by new Ph.D.s and graduate students in research groups.

Groves alone did not create the organization. At Los Alamos, Dr. J. Robert Oppenheimer forged a hard-working team out of brilliant scientists assembled from American and European universities. He encamped a group of academics used to the easy political and personal freedoms of college life behind fences and guards; he got them to accept, albeit grudgingly, a large degree of military control of their lives. At Chicago's Metallurgical Laboratory, Dr. Arthur Holly Compton worked with a temperamental but brilliant group of senior scientists, recent Ph.D.s, and graduate students to think through questions of the properties of fissionable uranium and plutonium and to do long-range planning about the future of atomic energy.

Graduate students in physics in the 1930s often engaged in hard physical labor, and the long-standing infusion of craft skills into the field of physics continued in a new fashion at Chicago. At Berkeley's Radiation Laboratory under the leadership of Ernest O. Lawrence, a whole generation of young physicists had spent much of their graduate and postdoctorate years working as electricians and mechanics on the assembly and operation of heavy electrical equipment in the cyclotrons there. At Chicago in 1942, physics graduate students would find themselves pitching into the grimy and back-breaking labor of constructing nuclear reactors out of stacked graphite blocks, learning and applying the skills of carpenters and masons.[10]

The Physics behind the Reactors

At the heart of the emerging culture and profession of nuclear engineering was a new type of machine, the atomic pile, or reactor. The conception of that machine derived from breakthroughs in science that had occurred in the late 1930s. The particular form and design of the machines evolved out of the organized efforts of several key scientists.

In 1939, published results of the findings of Otto Hahn and Lise Meitner in Germany indicated that atomic fission occurred in uranium atoms when uranium was bombarded by neutrons. In the United States, research confirming fission and work on the possibility of controlled nuclear fission in a chain reaction went forward at first informally among several groups, and then under the auspices of the federal government's Office of Scientific Research and Development (OSRD), headed by Vannevar Bush. In April 1941, Bush asked James Conant, president of Harvard University and chairman of the National Academy of Sciences, to prepare a report on the possibility of using nuclear fission to produce atomic bombs. Conant established the S-1 Committee, headed by Arthur Compton, to draft a report on the state of nuclear science. To conduct further research, Compton suggested, work under way at Columbia under Enrico Fermi should proceed. The next step would be the "production of chain reaction with carbon and uranium."[11]

Leo Szilard, in correspondence with Enrico Fermi at Columbia, had suggested in 1939 that if the neutrons emitted by U-235 could be slowed, or "moderated," by placing uranium in a lattice or structure built from a low-atomic-weight element, such as carbon, hydrogen, helium, or beryllium, a chain reaction could be started. The chance of a slowed neutron impacting another nucleus and producing another fission would be increased by the moderator. By changing the design and quality of the materials in the moderator-uranium lattice, the neutrons produced could be multiplied. If on average, each neutron generated more than one more neutron through impact and fission of U-235 atoms, the reaction continued; if each neutron generated, on average, less than one more neutron, the reaction would die out. The point of equilibrium would occur when a second generation of neutrons exactly equaled the first generation. A reaction with more neutrons in the second generation than in the first would continue.[12]

The physicists expressed the point of equilibrium as $k = 1.0$; reactions at

Enrico Fermi designed the world's first successful reactor, and he started it before a small audience on 2 December 1942 in the West Stands Unit of Stagg Field at the University of Chicago. (U.S. Department of Energy)

any number greater than 1.0 would continue, and those with k at less than 1.0 would die out. In July 1939, Szilard and Fermi agreed that carbon in the form of graphite bricks would be a readily available and useful material to serve as moderator and that a pile constructed of such bricks offered a good chance of demonstrating a chain reaction. At Columbia, and later at Chicago, Fermi conducted experiments with small "exponential piles" of graphite, measuring 5 to 10 feet on an edge, to estimate the rate of multiplication of neutrons and to estimate the point at which k would exceed 1.0.[13]

In this process some U-238 nuclei would acquire, or "capture," a neutron and decay through intermediate steps to element number 94, eventually called plutonium (Pu-239), which itself, the scientists predicted, would be fissionable. As U-235 or Pu-239 atoms fissioned into two approximately equal elements, some matter would be converted to energy. A controlled reaction would release its energy in the form of heat and radiation; an uncontrolled reaction with a critical mass of U-235 or of Pu-239 would constitute a bomb and release its energy in a nearly instantaneous burst of shock, heat, and radiation. Since there was plenty of U-238 and very little U-235 (the ratio of these two isotopes in natural, or unseparated, uranium is 99.3% U-238 to 0.7% U-235), a good approach to getting fissionable material would be to make plutonium out of the more plentiful U-238. The new element with the atomic weight of 239 could be produced in a slow thermal reactor. Hence the concept of a production pile.

As Fermi worked on the exponential piles, he discovered that impurities in the graphite would absorb neutrons and "give back none in return," tending to reduce the approach to k at 1.0. Fermi and Compton called those impurities "poisons" because they tended to kill the reaction; they understood the need for high-quality graphite blocks of great purity to achieve a sustained chain reaction. As early as April 1942, Compton became convinced that, with good-quality uranium and graphite, a self-sustaining chain reaction could be produced. He reported to the S-1 Committee that he believed a production pile could be built and that the plutonium produced could be separated from the uranium in the piles.[14]

Organization of the Metallurgical Laboratory

Early in 1942, Arthur Compton and groups from Columbia and Princeton consolidated research under OSRD at Chicago. OSRD contracted with the University of Chicago to fund the Metallurgical Laboratory, beginning the pattern of university operation by contract that came to characterize the research side of nuclear work in later years. The "Met Lab," as the Chicago group was called, was the first of what became the atomic establishment's system of national laboratories.

Within the Met Lab, Compton organized the scientists into groups or committees to specialize on different aspects of the problems. Fermi headed the physics group to design, build, and test the pile to demonstrate the chain reaction; Eugene Wigner headed the theory group to design the full-scale production pile to make Pu-239. Compton formed a group, the Engineering Council, to design a pilot production pile, as an intermediate stage between the demonstration pile and the full-scale pile. The Engineering Council consisted of physicists John A. Wheeler, Samuel K. Allison, Enrico Fermi, Norris Hilberry, Richard Doan, and Frank Spedding, together with a petroleum engineer, Thomas Moore, and chemists Glenn Seaborg and Miles Leverett. Moore chaired the council.[15]

In the language of management used decades later, conceptual design for the demonstration reactor, the intermediate or pilot-plant reactor, and the eventual production reactor had been assigned to three overlapping matrixed groups. Through this early period, the management that evolved under Compton's leadership reflected the easy-going, collegial style of academics who would be interested in and contribute to each other's work, rather than the more structured, compartmentalized, and centralized management style that the military would seek to bring to the effort later.[16]

During the summer of 1942, the various groups made rapid progress. The intermediate pile designed by Moore's Engineering Council took shape on paper, first as a graphite-moderated, helium-cooled pile with vertical columns containing uranium-graphite cartridges.[17] Wigner's group proceeded, again on paper, with a water-cooled design. Both helium and water cooling had disadvantages: helium would require new pump designs; water could lead to corrosion. Another group, under Leo Szilard, investigated the possibility of a liquid bismuth–cooled pile. By September 1942, no firm choice had been made among the methods of cooling.[18]

The Manhattan Engineer District and the Du Pont Contract

OSRD recognized that the scientists at the Met Lab simply did not have the organizational or business experience to take on the large-scale construction that would convert their discoveries into a system of industrial production. Moreover, overseeing the work of industrial corporations would require more expertise in construction, engineering, and contract management and supervision than OSRD could mount by itself for an operation of that magnitude. Therefore, in June 1942, OSRD began to shift the supervision of the project to the Army Corps of Engineers. During that summer, the Corps appointed Col. James C. Marshall from the Syracuse Engineer District to take charge.[19] Marshall worked closely with the Met Lab and selected the engineering firm of Stone and Webster as principal contractor to build the planned reactors. The scientists working under Compton rebelled in early September over the choice of Stone and Webster, most of them finding the company's representatives intellectually weak and insufficiently experienced. To coordinate and move the project along, Gen. Brehon Somervell of the Army Services of Supply appointed Col. Leslie Groves to head MED on 17 September. Groves had already established a reputation for getting a huge job done to a tight schedule because he had supervised construction of the largest building in America, the Army's new headquarters, the Pentagon. A week after his selection to head MED, Groves was promoted to the rank of brigadier general.[20]

Groves agreed with the scientists that Stone and Webster lacked the experience and commitment for the project and soon contacted executives at E. I. Du Pont de Nemours & Company to replace the earlier contractor. Some existing contracts with OSRD for work at the Met Lab continued, with the Army establishing new contracts as the old OSRD contracts phased out in April 1943. MED took over the system of OSRD contracts and expanded on it, establishing the patterns of mixed academic and industrial contracting that would survive into the postwar years.[21]

Groves instructed Compton to move along with pile design, if necessary following up on more than one design. As a consequence, the three design groups under Fermi, Moore, and Wigner went ahead; the liquid-bismuth design being considered by Szilard's group, since it depended on an exotic material, was treated as a remote future possibility, with less staff and funding.[22]

After they accepted the MED project, Du Pont officials sought to make clear that they had not eagerly pursued a central role in developing the nuclear weapon, hoping to avoid a repetition of the "merchants of death" criticism the company had endured in the post–World War I period, as part of the munitions industry.[23] For this reason, they carefully documented the stages of their involvement, later providing General Groves, at his request, with corrections to his draft of the Manhattan Project history. Du Pont executives and engineers pointed out that they had not entered the atomic bomb project for profit. Instead, they had accepted the work because Groves and a small circle of Roosevelt's advisers convinced the company of the national need.[24]

The Du Pont participants wanted to stress the "stepwise" involvement of the company, implicitly denying any corporate rush to get into the massive wartime project. The company's involvement began with some initial consulting work by a small team of Du Pont specialists, led by C. M. Cooper, in the summer of 1942. In October 1942, Du Pont accepted a letter contract to design a "semi-works separations plant" for material produced at the Met

Lab. Over the period 4–6 November, a three-man team from Du Pont, headed by Crawford Greenewalt, who was chemical director of Du Pont's Grasselli Chemicals Department, met with Arthur Compton and Norris Hilberry at Chicago. This team reported to Du Pont's executive committee later in November. Groves convinced W. S. Carpenter, Jr., president of the Du Pont Company, of the national significance of the project, and over the next weeks, Du Pont worked out a one-dollar fee contract for constructing the full-scale plant, finalizing a letter contract for that work on 21 December 1942.[25] Greenewalt was appointed manager of the Technical Division Explosives Department (TNX) and played a key role in reactor design decisions, serving as liaison between Du Pont's Wilmington, Delaware, office and the Chicago Met Lab.[26]

Although both the Army and Du Pont had used "cost-plus-a-fixed-fee" contracts before, the Du Pont letter contract and the final contract restating it had several notable ways of arranging the relationship between the government and the contractor. While Du Pont received only a one-dollar fee on what became a half-billion-dollar project, the company insisted on and obtained several clauses that came to characterize the operating contracts of the modern nuclear establishment.[27]

Under the contract, Du Pont could continue to apply corporate pay scales, rather than government pay scales, to the employees it transferred to or hired for the project. This was a significant and important concession, for corporate salaries at that time were in the range of 150 to 250% of the amounts paid for equivalent work to government technicians and engineers or to university faculty. The consultants who had worked at Chicago in the summer of 1942, including C. M. Cooper, had been "loaned" to the University of Chicago and placed on university salary. Du Pont noted that the university "could not extend to these employees salary treatment and benefits of the type provided by Du Pont industrial relations plans commensurate with their current status as Du Pont employees." Learning from this experience, Du Pont designed the letter contract and the final contract to permit the company to treat its transferred and hired personnel as Du Pont employees.[28]

The government also undertook to reimburse Du Pont for all costs and losses incurred as a result of the work, including normal expenses for administrative and general expenses allocated to the labor. Under the arrangement, the government protected Du Pont from losses which might result from the work; furthermore, the government took possession of all products. The possession clause was indeed crucial to Du Pont, as it turned out, for the major product, plutonium, as well as many by-products and wastes, were highly radioactive, dangerous, and long-lived. In a special clause, Du Pont retained the option of leaving the enterprise nine months after the war ended. Groves recognized that the cost-reimbursement, loss coverage, and exit clause aspects of the contract were unique, but he believed they worked to the government's advantage because they brought in a single large firm capable of the work. He convinced the government's comptroller general to accept the contract with its unique features.[29]

On the immediate level, the letter contract and the finalized contract allowed Du Pont to become involved. Without the protection for labor and administrative costs and the postwar escape clause, Du Pont simply would not have been able to get its best management employees to voluntarily work on the Hanford project. With much of the chemical firm devoted to war-related industry, Du Pont managers had many opportunities to serve the war effort through the firm. Du Pont management did not think it proper to ask long-term staff to sacrifice company rank, salary, and benefits to work as underpaid civil servants for the duration or longer.[30]

Stagg Field

Meanwhile, Chicago Pile 1, or CP-1, built under Fermi's direction, achieved criticality on 2 December 1942 and ran at slightly over k for a few minutes in the middle of the afternoon. At the time, this pile was called the West Stands unit because of its location under the university's football stands, or more simply, "the pile."[31] As a dramatic story, the events of 2 December were told and retold. The schedule of the construction of the reactor, the somewhat orchestrated moment of criticality, and perhaps an element of showmanship on Fermi's part contributed to those perceptions and those memories.

Fermi had originally planned to stack the graphite blocks to 76 layers. The crews learned that the easiest way to shape the blocks was with standard woodworking tools such as power saws, planers, and drills. The stacking work began 16 November 1942, with two 12-hour crews, one under the supervision of Walter Zinn and the other under Herbert Anderson, laying down three or four tiers of blocks a day, setting a predictable rate. Fermi modified the design as the pile was built because a combination of a better grade of graphite and a more refined uranium than originally planned allowed him to estimate that the pile would only need 57 layers of bricks. By late November, a simple calculation projected a completion date in early

December. Word of the anticipated date of completion spread among the scientists; the extended "stage wait" contributed to the sense of drama and history-in-the-making recorded by many of the participants.[32]

Fermi's crews inserted uranium metal and uranium oxide in roughly spherical shapes as fuel in some of the graphite blocks, alternated by layers of "dead" or solid graphite blocks with no fuel in them to provide the moderating effect. They ensured control of the reaction by leaving ten cadmium strips inserted in the pile; as a neutron absorber, cadmium would prevent criticality. These early "reactor control rods" were no more than 13-foot-long pieces of cadmium nailed to wooden strips and inserted in channels left in the pile of graphite blocks. Fermi, Zinn, and Anderson prevented accidental removal of the strips by padlocking them in place.

On the day of the crucial experiment, all but one of the control rods were removed. One of the removed rods, called "Zip," was held up by a solenoid mechanism, designed by Walter Zinn, that would release the rod automatically if the neutron flux exceeded a certain point or on electric command from one of Fermi's assistants. A second was tied off by a rope to the rail at the edge of the balcony overlooking the pile, where most of the observers gathered. Hilberry stood by with an axe, ready to sever the rope should the reaction get out of hand. The last control rod would be removed slowly, in 6-inch increments, to allow the pile gradually to approach $k = 1$ and then go to the self-sustaining level of $k > 1.0$. A special "suicide squad" of three young physicists stood on a platform above the pile with jugs of cadmium sulfate solution which would be dumped if all else failed to control the reaction. The presence of Zip, the axe, and the squad with the jugs all made the question of reactor risk visual, heightening the theatrical sense.[33]

On the morning of 2 December a quiet but excited crowd watched as Fermi calmly called for 6-inch incremental removals of the last control strip and kept his eye on a recording stylus on a drum of graph paper that indicated the radiation levels. He relieved the tension, yet also contributed to it, by breaking for lunch. After lunch he resumed his work, with Wigner, Crawford Greenewalt, and 39 others in attendance. At 3:49, Fermi announced, "The reaction is self-sustaining." The chain reaction had continued for less than 5 minutes when Fermi ordered the control rod reinserted.[34]

After the experiment, Compton called James B. Conant at Harvard. He had no prearranged code but wanted to pass the word of the accomplishment. His sense of the drama and history was conveyed in the impromptu communication.

"Jim, you'll be interested to know that the Italian navigator has just landed in the new world," said Compton. To explain that the work on the pile had gone more quickly than anticipated, he added: "The earth was not as large as he had estimated and he arrived at the new world sooner than he had expected."

"Is that so?" Conant replied. "Were the natives friendly?"

"Everyone landed safe and happy," replied Compton.[35]

The first nuclear reactor, from which all others can be said to descend, had many of the elements, in a crude form, that came to characterize its descendants: a moderator in the form of graphite and a passive air cooling system; control rods and emergency safety systems; monitoring and recording devices; and fuel in the form of balls of uranium metal and uranium oxide. In a sense, there was even an evacuation plan: Fermi said if the reaction failed, he would walk away.[36] Fermi performed the roles of reactor designer, construction manager, and control room supervisor.

None of those roles and none of the lexicon was quite so formalized; the vocabulary was being assembled along with the device. Early in his studies of the principles involved, Greenewalt learned about carbon as a moderator. He called it a "slow-downer," a term which—perhaps fortunately—did not survive in nuclear jargon. A red button switch controlling Zip for rapid insertion was labeled, almost as a joke, "scram." That word, of course, became part of the language, both as a verb and a noun in the nuclear world; at the time, it reflected the evacuation plan of those less confident than Fermi.[37]

The dramatic demonstration had immediate organizational and technological consequences. Greenewalt's presence was perhaps a lucky accident, perhaps part of the planning by Fermi and Compton, to swing Du Pont into line. Du Pont's liaison committee, which had expressed some earlier skepticism about the project, was now headed by a witness to the historic moment. At the surface technical level, Fermi's pile, by proving that a controlled reaction could work with graphite moderation, advanced the prospects of the various graphite designs.

On the technical side, the demonstration had several other long-range consequences. The larger-scale production reactors would run hot for sustained periods and, unlike CP-1, would have to be positively cooled. A water-cooled reactor, which prior to Fermi's experiments had seemed like a remote possibility because the water coolant would absorb some of the neutrons, now seemed within the realm of feasibility, since Fermi had found k easier to achieve than anticipated. As the helium design group studied Fermi's results, they considered the possibility that air cooling might also work, avoiding some of the technical problems encountered in working with helium.[38]

Du Pont and the Selection of the Hanford Site

Following reports of Fermi's success, the Du Pont executive committee decided, Greenewalt noted in his diary with a sense of excitement, to "take on the 'Chicago' project, lock stock and barrel—or in other words design, construction and operation." That afternoon, Greenewalt heard he would continue to be involved in some capacity. The next day, Du Pont executives convinced General Groves that Greenewalt should remain in Du Pont's personnel "setup," referring to the letter contract that confirmed that Du Pont employees could retain their salaries and benefits.[39]

It soon became apparent to Greenewalt that he brought a point of view that was different from that of the Chicago group of academic scientists. Greenewalt found Compton's organization weak in management at the top, and he thought Compton's views on the differences between scientific and industrial work "peculiar." In particular, Greenewalt disagreed with Compton's plans for engineering the full-scale production pile. Greenewalt thought the Chicago group ought to be "small and consulting rather than experimental."[40]

Greenewalt found the scientists in Chicago difficult in several other ways. He thought Leo Szilard "a queer fish" and believed he had to reassure Fermi and Eugene Wigner that he would keep them involved in the planning for the production pile. He saw they felt "very keenly" the importance of keeping in close contact, fearing that otherwise some small design detail "might violate physical principles."[41]

Over the last two weeks of 1942, Du Pont cooperated with the Corps of Engineers in selecting site "W," at Hanford, for the first production reactors, following eight criteria established by Groves in consultation with Chicago scientists and Du Pont engineers. The requirements were very specific: 25,000 gallons of water per minute; 100,000 kilowatts (kW) of available power; a rectangle of approximately 12 by 16 miles for the hazardous manufacturing area; a laboratory area located at least 8 miles from the nearest pile or separations plant; an employee village located no closer than 10 miles upwind of the nearest pile or separation area; at least 20 miles between the piles and separations areas and the nearest existing community of 1,000 or more inhabitants; no railroad or main highway closer than 10 miles from the piles and separations areas; and a climate that would not affect the process.[42]

The company appointed Du Pont engineers A. E. S. Hall and G. P. Church to explore sites on the same day Greenewalt was appointed. General Groves sent Hall and Church with Lt. Col. Franklin T. Matthias, who later served as the Corps' area engineer supervising the project. The three-man team first met with Groves to review the site requirements, then examined on paper a series of 20 sites conforming to the eight criteria, selected from map review by the Corps of Engineers. Groves made it clear to the team that he had thought about the sites and preferred the Pacific Northwest.[43]

Hall, Church, and Matthias spent the Christmas week visiting sites in the West, starting from Seattle and Spokane. Working their way generally southward by air, they visited or flew over five relatively promising sites: Coulee and Hanford in Washington State and Pit River, Needles, and Blythe, in California. By 2 January 1943, Hall and Church had prepared their report, recommending Hanford.[44] On paper, all five sites seemed to come close to the requirements of an available large tract with isolation from population, low land costs, available power in the range of 100,000 kW, and available water supply. But all of the sites except Hanford had several specific disadvantages: the Coulee site would require 23 miles of pipeline for water and the land value was moderately high; the Pit River area had high land values and would require relocation of railroads and highways; the Needles site was in an earthquake zone, would require relocation of a highway, and suffered extreme summer heat; that the Blythe site was within 50 miles of the Mexican border precluded it from consideration on security grounds. The only disadvantage at Hanford was the lack of natural camouflage due to the flat, sagebrush-covered land. The low land value at Hanford was an advantage. After Church and Hall's recommendation of Hanford, Matthias requested and received prompt evaluations of it and the other three possible sites from the Corps of Engineers' power consultant, who also favored Hanford over the other sites.[45]

Never again in the history of nuclear reactor siting in the United States, did planners reach such a major decision so quickly. However, the haste of the decision did not mean a disregard for safety: it is also notable that, in later years, no site chosen for any commercial nuclear reactor had the same extent of raw, uninhabited land serving to insulate the general population and the facility employees from the risk of radiological exposure. Hanford was chosen on the basis of safety, security, economic, and utility criteria.

Wartime secrecy precluded seeking the opinion of, or asking the consent of, the local population, the state government, or even the state's congressional delegation. None were even informed of the proposed use of the huge federal land area acquired.

Corps appraisers filed reports on 21 and 23 January, and the Corps began formal acquisition of the land on 9 February. Despite the generally arid nature of the landscape, condemnation and buyout took more time and money than Groves would have preferred; he grumbled at the small inconvenience of dealing with the civil courts and the local population. Eventually, the Army acquired an area half the size of Rhode Island for the reactors and their associated support and separations facilities.[46]

With Du Pont lined up and the site chosen, Groves and Greenewalt faced the question of exactly what type of reactor to build. Over the next months, they worked with the Met Lab scientists in sorting through the alternate conceptual designs and moved quickly to a commitment to one type of reactor for Hanford.

2 • Building Reactors at Hanford

Under Gen. Leslie R. Groves's leadership, design choices and construction followed rapidly on site selection. Du Pont's Crawford Greenewalt worked out relations with the group of scientists at Chicago, settled on details of the design, and started construction. By September 1944, Du Pont had the first of the three wartime production reactors, B, in operation, with reactors D and F following within a few months. Du Pont also arranged the construction of the X-10 reactor at Oak Ridge, completed during 1943–44. The specific form and shape of the Hanford group of reactors and the Oak Ridge reactor resulted from a series of rapid decisions. The specific technical choices represented hasty coordination and compromise between the academics at Chicago, the industrial engineers under Greenewalt, and Groves's army officers in the emerging MED culture.

The first weeks and months of operation of the Hanford reactors demonstrated that the concept of a nuclear reactor could lead to industrial-scale machines that could be run as factories. Furthermore, the successful production of plutonium not only made possible the nuclear weapon used at Nagasaki but also laid down part of the organizational and cultural basis for the postwar nuclear weapons complex and demonstrated emerging patterns of technical decision-making.

Conceptual Design and Relations between Contractors

While the selection of Hanford as a site went forward, Greenewalt worked on organizing his team and clarifying the role of Du Pont. He proceeded rapidly with recruiting, orienting new staff, and assigning responsibilities during December 1942 through January

1943. As he discussed progress with the scientists at Chicago, he grew increasingly frustrated by their plans for the production pile and by their lack of structured organization. On 28 December 1942 he noted that despite some preliminary thinking, there was "no mechanism yet devised for unloading and sorting, no flow sheet, operating manual or program. No clear idea as to what Du Pont is expected to do—Hell!" He approached the lack of design decisions as a manager: "The first thing to do is to work out an operating organization."[1]

Greenewalt recognized that he was joining a going organization and believed it essential to "infilter" the pile design group, "in spite of the fact that we aren't very welcome." He tended to agree when his engineering staff members complained of being "not properly used and too much under domination of the physicists."[2]

As he struggled with these organizational issues, he also dealt with the major issue of the conceptual design phase: whether the proposed full-scale production reactor should be air-cooled, helium-cooled, or water-cooled. Water cooling seemed dependable but had disadvantages in that neutron absorption by the required aluminum coatings and the moderating effect of the water would both reduce reactivity. Helium cooling presented other difficulties: the need for new pump designs and the problem of working with an unfamiliar material. Air cooling would not effectively transfer the expected high heat of a full-scale production pile.[3]

X-10: Reactor for the Separations Pilot Plant

As the Du Pont engineers took over planning for the full-scale plant, they also agreed to assist in the construction of a semiworks, or pilot plant, for the separation of plutonium from the irradiated fuel slugs. In connection with the semiworks, they agreed to construct X-10, a pilot pile to produce small quantities of irradiated slugs for use in separation experiments in the semiworks.

The designation of the X-10 reactor as the pilot pile led to a misunderstanding among some in the project that the reactor *itself* had been intended as a pilot plant for the Hanford reactors. The fact that X-10 was to be air-cooled and the Hanford reactors water-cooled suggested that X-10 would be inappropriate as a scale-up model for the Hanford reactors, although the difference presented no problem if X-10 was to serve as a supplier for the separations semiworks. The story of how X-10 came to be designed with air cooling, while the production reactors followed a separate design, reflects

Du Pont's willingness to take over and make use of the Met Lab's existing committee approach in the early stages.

On 31 December 1942 Greenewalt wrote that he did not object to Du Pont's role in building X-10 and the pilot separations plant, but he did not want the company involved in their operation because of the hazard and liability issues. Instead, if Du Pont staff could get operations training under the responsibility of the Met Lab, "we get what we want and duck liability." At this stage, Greenewalt saw the proposed semiworks plant as "a wonderful opportunity for pilot plant testing and later for operator training and instruction." Five days later, Groves signaled his agreement by issuing a letter contract to Du Pont to construct the reactor for the pilot plant early in January 1943.[4]

In January, Greenewalt and Roger Williams at Du Pont reviewed the safety issues involved in siting the pilot plant reactor in Argonne Forest, 20 miles west of downtown Chicago, as had been the original plan. Greenewalt obtained the population figures: within a 1-mile radius there were 100 people, and within a 5-mile radius, there were 8,750. Roger Williams had consulted with John Wheeler, the Chicago physicist who was assigned to work regularly with Du Pont. According to Wheeler, an accident involving vaporization of the uranium fuel would deposit lethal radiation to a 5-mile radius. Greenewalt and Williams concluded that the risk at the Argonne Forest site was too great and decided to move the pilot pile to "site X," a 59,000-acre site near Clinton, Tennessee, later renamed Oak Ridge, which Groves had acquired in September 1942 for uranium separation facilities. Here, ridges that inhibited prevailing winds and some isolation would insulate the proposed reactor as well as the separations plant from surrounding communities.[5]

Greenewalt anticipated that making the decision to relocate the pilot plant reactor to site X without consulting Chicago Metallurgical Laboratory director Arthur Compton would be a mistake. The move to Clinton would reduce from 11 months to 9 months the amount of time that the separations pilot plant could be operated prior to the scheduled opening of the first Hanford production plant. While Greenewalt did not see the delay as too serious, he felt that the decision would be a "blow to the Chicago group," particularly because Compton had stated that Argonne was safe. Greenewalt anticipated "hard feelings," since it was "a nasty situation badly handled." After Groves met with Compton, Compton agreed to transfer the reactor to site X, but the decision caused Compton considerable heartache.[6]

Du Pont's stepwise involvement, together with the fact that two Chicago design groups had already started planning piles, led to a quite different basic design for the X-10 reactor from that of the full-scale reactor. For the Hanford reactors, Du Pont had briefly explored the helium alternative, which would require extensive development of pumps and which made Greenewalt "gloomy" to think about.[7] Three of the men from the early technical group that had considered the helium alternative in the summer of 1942—Moore, Whitaker, and Wheeler—formed the core of the group working with Du Pont to design the X-10 reactor and they followed the helium design, substituting air for helium.[8] On 16 February 1943 Williams at Du Pont decided on the general configuration of the Clinton X-10 pile. It was to be a cube 24 feet on a side, sitting on one face, with horizontal 1.1-inch rods, 8 inches apart, center to center. On the same day, Greenewalt accepted Wigner's water-cooled concept for the Hanford production piles, clearly dropping consideration of helium or any other gas as a coolant for the production plants. In effect, on 16 February, confronted with lots of hard work by two different Chicago-led groups, Du Pont used the plans of each, for two different reactors.[9]

In 1945, as Groves assembled material for a thoroughly documented history of the project, Roger Williams of Du Pont noted that the main function of X-10 had been to provide plutonium for pilot-plant separations work at Oak Ridge, reflecting its original purpose. He complained that a myth had grown up that Clinton had served as a model for Hanford as a result of "a widely held misconception of the purpose, limitations and contribution of the Clinton semi-works." Williams wanted the official history to say that "the Hanford production units . . . had to be designed, constructed and operated without major guidance from Clinton experience." Williams was correct.[10]

As soon as the smaller X-10 reactor began to operate in December 1943, its assigned tasks reflected that its mission was very close to what Williams recalled, rather than what the "myth" suggested. Under the general administration of Compton from Chicago, the local management fell to M. D. Whitaker, director of the Clinton Laboratories (as the facilities in Oak Ridge were now called), and to R. L. Doan, coordinator of research at Clinton.[11] At the startup of the reactor, Compton forwarded a detailed mission statement for X-10 to Whitaker. X-10 was to have a technical program, a training function, and responsibility for production of experimental quantities of product. The production of small quantities of plutonium for separations experiments and for use at Los Alamos was the most urgent of the several overlapping missions. Under its technical program, Clinton was to proceed with studying methods of separating plutonium from the uranium fuel elements, working with both a bismuth-phosphate separation plant and an alternative lanthanum-fluoride process. From time to time, special Hanford-related studies would be requested. The mission statement contained nothing to suggest that the X-10 reactor would serve as a model for the Hanford reactors.[12]

By January 1944, Whitaker had followed Compton's program for X-10, assigning personnel and time in specific proportions: 12% to product production; 75% to product isolation; 4% to product utilization; 9% to health protection. Of the 32 listed assignments as of January 1944, only three projects related to Hanford reactor design: a study of waste-handling procedures at both Hanford and Clinton, a test of corrosion of Hanford-style aluminum tubes and slugs, and an evaluation of shielding to be used at Hanford.[13]

The X-10 reactor also served the Hanford operation as a training ground through 1944. Two groups of Du Pont employees, a total of 183, went through the Clinton "school" before moving on to Hanford. A group of 29 Clinton employees, mostly specializing in health hazards, also trained on X-10 before moving out to the production reactors at Hanford.[14]

The Clinton reactor's relationship to the production reactors grew even more tenuous as organizational changes continued. Although Du Pont built X-10, Groves put its operating management in the hands of the Chicago Met Lab. From the beginning, Compton and the others at Chicago were uncomfortable with the arrangement. On 1 July 1945, Chicago turned the operation of the Clinton Laboratories (including the X-10 facility there) over to Monsanto Corporation, which kept Martin Whitaker on as the director of the Laboratory.[15]

Environmental Impact and Mitigation

At Hanford, Du Pont and the Army rapidly addressed a series of issues that later generations characterized as environmental impact and mitigation of impacts. As soon as the site was selected, Greenewalt consulted meteorological studies to determine what would happen if "a pile blew up" during a weather inversion. Preliminary calculations indicated that emissions of radioactive xenon from regular operations would dissipate harmlessly but that a "bottleneck" of radioactive gas due to inversion could endanger nearby Pasco. Greenewalt requested more meteorological data before specifying exactly where to build the piles.[16]

In another case of concern for environmental safety and impact, Greenewalt examined the issues of radiation tolerance levels in water used for drinking, bathing, and uptake by fish. While he had sufficient data on drinking water, more study was needed on the issues of whole-body radiation resulting from human immersion and eating contaminated fish. He considered these issues very early on, taking them to the policy group in Chicago for further discussion. Considering the possibility that there would be "fission product leakage into effluent," he thought it possible to build retention basins for decay. That was the method eventually employed.[17] Work on fish research went forward under Army auspices, particularly regarding the effect of radiation on salmon.[18]

By May 1943, Greenewalt had a variety of groups reporting to him on different full-scale production pile design problems, some of which reflected environmental and safety concerns: shielding, control, water flow control, loading and unloading devices, coating of the uranium fuel, and water purification. On issue after issue, the Du Pont engineers moved to "freeze" the design of one component so that other elements could be designed and then built, on the assumption that the design of the previous one had already been set. The design of the final reactor thus emerged in stages, leading to a nearly identical design for all three reactors. Each was a cubelike structure about 34 feet by 46 feet by 41 feet high. The interior block of graphite that was the reactor core measured 31 feet by 40 feet by 35 feet. In consultation with Enrico Fermi on 28 May, Greenewalt decided to surround the block with laminated walls of steel and masonite, to make up the radiological, or "biological," shielding.[19] A cast-iron thermal shielding, 10 inches thick on the top, bottom, front and rear sides, and 8 inches thick on the right and left sides, surrounded the piles between the interior graphite block and the external laminated biological shielding. Du Pont also "froze," or set for planning purposes, the design level of power at 250 megawatts (MW), far exceeding the kilowatt level at CP-1 and the 40-MW level at X-10.

The front of the block was the charging face, the rear the discharge face. A total of 208 cooling water pipes ran front to rear, while 29 control rod holes punctured the pile vertically, together with 9 horizontal control rod holes from left to right. On both the front and rear faces, an elevator serviced the pile. In the front, the elevator supported a machine for charging the pile with fresh fuel; the rear elevator contained a cab for meeting emergencies arising from stuck discharge elements. The fuel was sealed in aluminum "cans," or "slugs," reducing the likelihood of uranium fuel entering the water coolant, which circulated directly around the slugs. Each loading tube would be monitored for radiation in the water, with a separate "Panelit" gauge in the control room for each of the 2,004 tubes, so that a ruptured slug could be immediately detected. In the reactors, minute amounts of U-238 would be converted by the addition of neutrons and after a process of decay to Pu-239, that material would have to be refined out from the discharged slugs. The anticipated rate of production was very low: tons of discharged slugs would yield pounds of final product. Over objections from Chicago, Greenewalt insisted on building in several hundred excess process tubes to allow for unforeseen needs.[20]

Through May and June, Greenewalt discussed the crucial questions relating to water cooling with his designers, participating in decisions that narrowed the design choices. Early in May, when production estimates still suggested that four reactors would be required, Du Pont engineers settled on demineralizing and refrigerating the incoming water for two of the piles and using raw water for the other two. Greenewalt, who had taken a short vacation, independently developed a similar concept.[21] The final decision on this matter was to install water treatment plants for the now three reactors, but refrigeration only for the last two (D and F) of the three built.[22]

By 2 June, Greenewalt and his colleagues had settled on calling the system a "once-through" cooling system, meaning that heat exchangers would not be used to reduce the temperature of the water after it exited the reactors. The fuel slugs would be canned so that the radioactive fuel would not get into the coolant except from an accidentally ruptured can. Nevertheless, radioactive fission products would escape into the coolant during routine operation. In order to mitigate the effect of the heat and of radioactive products on the river, designers set up a system of cooling ponds and a venturi design at the outlet from the pond to the river, which would dilute the effluent to a one-part-in-ten ratio with river water. To some extent, fission products would decay in the cooling ponds; then, they would be released. The impact of the effluent water on fish, Greenewalt concluded, would be none as long as the water was detained for an 8-hour period after cooling. Inevitably, the Columbia River would show increases in radioactivity; the level had to be held below thresholds dangerous for human or fish uptake.[23]

Early in 1943, in a separate effort to address somewhat similar environmental concerns, Compton recommended to Groves that CP-1 be dismantled and moved out of downtown Chicago to a site in Cook County, near the Argonne Forest, where it would be rebuilt in a structure of its own, block by block. His recommendation was implemented, and the pile was redesignated CP-2. It was intermittently run at about 100 kW, much higher than had been deemed safe in the Chicago location. Until then, the pile had been formally called the West Stands Unit; after it had been moved, the practice of designating the major reactors operated by the Met Lab as a "CP" series began.[24]

The Heavy-Water Alternative: Wigner vs. Du Pont

As Du Pont went forward with the design for both the Clinton and Hanford piles in the early months of 1943, more and more decisions shifted from the hands of the scientists into the hands of Greenewalt and Williams at Du Pont. This decrease in responsibility did not sit well with some of the scientists, most notably Leo Szilard and Eugene Wigner, although Fermi, Henry Smyth, and others also had complaints. Wigner felt particularly bypassed with the Du Pont arrangement. His grievances included charges that Du Pont was attempting to monopolize the nucleonics industry, that Du Pont engineers were stalling work on a heavy-water-moderated design, that they delayed choosing water cooling unnecessarily by 2 months, and that they refused to take his advice about design. Charge by charge, Compton answered Wigner's complaints. "The fact is," Compton told Wigner, "that your antagonism to Du Pont is based upon beliefs which I know to be false."[25]

Wigner pointed out that others shared some of his views and then offered or threatened to resign if he had to continue working with Du Pont.[26] Compton diplomatically assured Greenewalt that Du Pont engineers should not feel too bad about Wigner's remarks about resigning; apparently Compton dealt with such threats routinely.[27]

Further tensions developed in Chicago when Wigner's group took up the possibility of a heavy-water-moderated pile. This project, labeled P-9, held great promise as a backup in case the graphite piles developed major problems in practice. However, since the heavy-water pile would require design effort and management attention, Greenewalt and the Du Pont engineers argued for a slow approach to the problem. A zero-power heavy-water pile should be built at Argonne, they argued, with a later scaleup possibly scheduled for Clinton. Greenewalt believed the heavy-water pile should be "homogeneous"—that is, that the heavy water should be used both as moderator and as coolant and that it should carry the uranium as fuel and target in a slurry, so as to be as different as possible from the graphite-moderated, light-water-cooled, solid-slug-fueled Hanford piles.[28]

In June 1943, Du Pont engineers urged that production of heavy water for a possible heavy-water-moderated pile be held off until the operation of a graphite pile could be "more clearly appraised."[29] Wigner and his colleagues already felt somewhat distressed that Du Pont had delayed accepting their judgment on water cooling and believed themselves cut out of the practical design decisions regarding the Hanford reactors. Because of those tensions, Wigner found Du Pont's opposition to moving ahead immediately with the heavy-water alternative an added insult. Wigner particularly, but some of the rest of the Chicago group as well, saw the issue as one of a business firm, headed by engineers, taking over from the nuclear physicists. Science, they believed, was not being given its due. In August 1943, Groves ordered several committee hearings to investigate disagreements between Chicago and Du Pont over how much effort should be put into the heavy-water design. The meetings provided an outlet for the discontents of Wigner and some of his Chicago colleagues and incidentally served as a forum for the academic-industrial conflict.[30]

Groves set an extensive agenda for the P-9 committee, starting with issues related to heavy water design. He included questions of where the work should be done, its relationship to Canadian work, and the ideal scale of the work. Groves asked whether an experimental, or "Fermi," pile should be built and whether or not a semiworks should be constructed. He also asked whether or not work on a full-scale-production heavy-water moderated reactor should move forward and whether it would represent part of the total production picture or simply serve as insurance in case the graphite reactors planned for Hanford failed. Further, he wanted to know what sort of contractor would be ideal for heavy-water work. In all, he asked 33 questions about the proposed heavy-water alternative, providing an agenda for discussion of a wide range of policy issues about design, the design process, and the relationship between the science contractor (the Met Lab) and the engineering contractor (Du Pont).[31]

The committee heard reports on the progress at Hanford from Roger Williams of Du Pont and also received an analysis of the prospects of the Hanford water-cooled, graphite-moderated reactors from Columbia University physicist Harold Urey, the scientist who had first identified deuterium in 1934. Urey held that CP-1, which he called "Fermi's pile," had demonstrated that graphite moderation worked and that for that reason it had been correct to go ahead in January and February 1943 with a graphite-based design, rather than one based on heavy water. Urey took the position that the committee should recommend work on heavy-water moderation as insurance against failure of the graphite piles. He was concerned at that point about the effects of water corrosion at the Hanford piles and believed heavy water a much better alternative, particularly if the reactor was to be homogeneous in design.[32]

Greenewalt and Williams testified about their design choice for Hanford: "We are sufficiently confident of the success of the graphite pile that we would object strenuously to setting aside essential experimental and theoretical effort in favor of work directed toward a second line of defense."[33] Instead, they suggested that only an experimental heavy-water pile be constructed. The Du Pont culture called for making firm decisions about design and then moving along to invest time and effort on the assumption that the decision was frozen. Water cooling had been a basic early design decision. It would not make sense, given Du Pont's method, to abandon a choice once made, for it would mean abandoning subsequent work based on that decision. Compton also agreed that the prospects for the graphite, water-cooled design of Du Pont looked good.[34]

Wigner was less conciliatory. As he testified before the committee, he pointed out that the morale of the Chicago group would be improved if a new engineering company were introduced. He would prefer one that "collaborated more completely" and shared its responsibility "more evenly with the Chicago group."[35] He added that he thought collaboration with Du Pont had been "very poor," and he thought "many people in the laboratory are angry with them." Wigner estimated four months had been lost by Du Pont; he felt that the company had been "put in charge" and, after that, did not cooperate.[36]

Whitaker, who had been the Chicago scientist in charge at Oak Ridge, countered by stating that the cooperation with Du Pont there had been good, with few misunderstandings. If work on heavy water was to go forward, Whitaker argued for using Du Pont, since its staff had already "got their feet wet."[37] Compton thought a company other than Du Pont would avoid the charge of "monopoly" and that a digression by Du Pont into heavy-water work would slow the company's progress with the graphite approach. Yet he doubted if there were enough resources in manpower in the nation to support a full-scale heavy-water effort. Ultimately, Compton's arguments about the manpower requirements appeared to carry weight with Groves, who decided not to pursue the design.[38]

Szilard used the P-9 committee to air grievances about the method by which decisions were reached; he believed "compartmentalization of information [was] an indignity to scientists."[39] The P-9 committee heard a variety of opinions on Wigner's offer to depart from the project with the theoretical group, including suggestions from both Fermi and Greenewalt, who thought it a poor idea.[40]

As a result of the P-9 investigation, the committee recommended work on heavy water as "insurance" against failure of the graphite-moderated, water-cooled approach. However, they regarded the heavy-water work as a second priority, compared to the graphite piles. They proposed two heavy-water reactors, one of 100 to 250 kW, and another at higher power, using U-235-enriched fuel. An intermediate pile, of 40,000 kW, could be built at Clinton; and full-scale production reactors, of 125,000 and at 600,000 kW could be built at Hanford. Diplomatically, the P-9 group stated that Du Pont had done a fine job that could not have been surpassed by any other organization, but recommended that another contractor be brought in for design and construction of the new projects. The engineering, the committee recommended, should go forward at the University of Chicago and at Columbia University.[41]

From this shopping list of recommendations, Groves accepted only the experimental heavy-water pile design to be built at Argonne, which eventually emerged as CP-3. In effect, he accepted Du Pont's position, validating its recommendation against a full-scale heavy-water effort, rather than the position of the P-9 committee. After the war, CP-5, a 2,000-kW enriched uranium research reactor with a heavy-water moderator and a graphite shield, was eventually constructed at Argonne, representing a scaled-down version of the second stage of the P-9 committee's recommendation.[42]

Perhaps relieved that his complaints had gone on record, Wigner and his theoretical group stayed with the Met Lab. After working with the first heavy-water pile at Argonne, Fermi moved on to Los Alamos.

Construction at Hanford

Du Pont and the Army Corps of Engineers were well committed at Hanford by August, when the P-9 report was filed. For Groves, speed of completion was crucial. In order to affect the outcome of the war, a deliverable weapon was needed by late 1944 or early 1945. At the same time, things had to work. Thus, each design decision placed the engineers in a dilemma: cut corners to speed up work and endanger the chances of a working design, or scrupulously adhere to exacting specifications and possibly delay the work past the point at which a bomb might be useful in the war. The engineers took a stricter view of this matter than the scientists, again following Du Pont's corporate culture in this regard. Matthias, the Corps of Engineers' supervising engineer at Hanford, told Groves that there were five different areas where attempts to achieve close tolerances might slow construction: graphite block machining, base plates for the shield, base blocks for the pile itself, laminated pile at the charge and discharge ends, and clearances between the graphite and the cast iron blocks at the sides. Matthias favored sticking with the accurate tolerances rather than sacrificing them for speed.[43]

Groves sought an outside opinion on the matter; and R. C. Tolman, vice chairman of the National Defense Research Council, reported on the issue of the close tolerances on the shielding blocks. In the opinion of the Chicago scientists Fermi and Wigner, the Du Pont engineers were too strict in adherence to design, slowing the work. On the other hand, Tolman found, the Du Pont people maintained excellent records, showing a number of points where they had relaxed tolerances in favor of speed. Tolman recommended that Du Pont look for further opportunities to avoid "bottlenecks."[44]

Greenewalt went ahead without specific design input from the Chicago group on all details, making decisions with Roger Williams as to control, "last-ditch" safety systems, loading and unloading procedures, cooling, shielding, locations, canning, and materials handling. Detailed design work proceeded on the three reactors at once, with construction of ancillary buildings, separations areas, and other projects moving along at the same time. By March 1944, the plans were shaping up, with B reactor scheduled for completion in August, and D and F reactors to follow.[45] In general, construction went slightly faster on D and F reactors because of lessons learned and problems solved on B reactor during construction and because of variations in labor allocations to the various projects. Du Pont brought B to criticality in September 1944, slightly behind schedule, with D in December 1944, and F reactor in February 1945, both slightly ahead of schedule.[46]

Construction brought problems that had not been resolved during design. The graphite blocks had to be machined to 40-inch lengths, 5 by 5 inches in cross section. During construction, constant vacuum cleaning of graphite dust kept the dirt to a minimum. The issue of precision arose because of the need to align the various slots and holes for the fuel and water and for the control rods, and to minimize the accumulation of error across the large dimensions of the pile. At the anticipated temperatures, ambient oxygen could cause the graphite to ignite. It was not enough to evacuate the air; in order to prevent minute quantities of remaining air from poisoning the reaction or contributing to combustion, the air was to be replaced with pressurized helium. On the whole, graphite blocks and their various holes and slots were held to a tolerance level of 0.005 inch. The blocks were machined on the spot, in a separate building, and carefully remachined to specification with an identifying number on each block. The milling and drilling equipment was simple woodworking equipment, and accuracy was achieved by using preset jigs to hold the worked block in place, following the craft practices established in academic labs and at the CP-1 project.[47]

Hanford Operations

The startup of the reactors, beginning with B reactor, was a gradual process, with testing for helium and water leaks, repairing of the effluent 48-inch "sewer pipe," and checking intake strainers. All the horizontal and vertical control rods were checked, and the vertical rods were tripped simultaneously to ensure that they would work in an emergency. Safety circuits, instruments, ventilating fans, elevators, and smaller systems were all given a final check before pile charging began. The charging of the tubes was itself an experimental process, with central tubes in the core charged first, without water cooling and then later, tubes around the periphery charged to gradually build up to and over reactivity, or $k = 1$.[48]

In a moment only slightly less dramatic than Fermi's startup of CP-1 some 21 months before, B first went critical at 10:48 A.M. on 26 September 1944. Fermi was present at the B reactor startup and provided advice and "specific verbal approval" for a number of variations from the preplanned procedures. The plan was to move forward experimentally, loading tubes to intermediate power levels, checking performance at each level, and then moving gradually to the next level.[49]

Shortly after midnight, the reactor was stabilized at the 200-kW power level. However, as the reactor was moved to its next level, of 9 MW, a sharp decline in reactivity was noticed, and the reaction ran to a stop. At first, the operators assumed the effect might be due to leakage of boron solution from safety rods into the reactor, but a check revealed no such leakage. Someone suggested that the timing of the delay might indicate a radioactive decay element of one of the fission products and intuited a "poisoning" effect due to the presence of xenon-135, which was estimated to have a half-life of about 9 hours.[50] The xenon would present a large cross section and thus have an absorbing effect, but as it decayed, the effect would level off. To take care of that effect, however, would require loading the reactor with more fuel. Colonel Matthias, who was absorbed in the details of construction, labor arrangements, and the work progress on D and F reactors, was dismayed to learn of the effect at B. On 29 September, he flew to San Francisco to inform General Groves that some "unknown" fission product was causing the reactor to shut down. Groves was concerned and called Chicago at once to get the involvement of the Met Lab scientists in helping to identify the problem.[51]

Groves angrily asked why the problem had not been discovered earlier at the Chicago and Clinton reactors. At Clinton, the records of X-10 reactor were closely reviewed, and Compton wrote to Groves explaining how it was that the xenon "poisoning" had not been anticipated there. What he called the "Hanford effect" or the "oscillation effect of W pile" was hard to detect at Clinton because temperature variations had masked the xenon poisoning. Responding on the spot to the issue, Compton noted that with some redesign, the B pile might eventually be able to operate at 200 MW.[52] The same day, Walter Zinn tried to achieve the poisoning effect on the experimental P-9, the heavy-water reactor at Argonne (CP-3). Zinn found the effect at high intensity runs and, through examining the period of the interference with the reactivity, and determined there was little doubt that the poisoning came with xenon-135. In a sense, Zinn confirmed and graphed the notion suggested at Hanford a few days before.[53]

Had the poisoning effect been determined on the spot by the Hanford operators, or had it been discovered at Chicago, by a scientific experiment? While the priority of discovery in the xenon story became a bone of contention later between the "scientific" and the "engineering" camps, a close examination of the records suggest that both approaches and both kinds of people were involved. Captain Valente, who maintained a detailed day-to-day diary at Hanford, indicated on 27 September that six different possibilities existed: loss of gas (helium) and replacement by air; gas moisture; leakage of solution from safety devices; deposition of chromium on the aluminum jackets of the slugs; leakage of cooling water into the graphite; and varying water pressures. Corrective measures or tests eliminated these possibilities, and on 28 September, the effect was repeated. On the twenty-ninth, the levels and decrease were recorded carefully. On the thirtieth, Valente noted: "A proposed theory suggests that the pile is producing a self-poisoning agent,—a granddaughter of some fission product." He did not suggest

Placed in operation in September 1944, with cooling water from the Columbia River, and with an impressive array of support facilities, Hanford's B reactor underwent incremental upgrades from its original design level of 250 MW. (Westinghouse Hanford Company)

who proposed the theory, but it was clear that the suggestion was local, rather than from Chicago. By 3 October, the day Zinn filed his report, Valente referred to the poison as xenon-135. Zinn and Compton both sent memos on 3 October regarding the effect. The diary sequences support the view that the solution was identified first on the spot, then confirmed in Chicago.[54]

At Hanford, teams began loading additional tubes, up to 1,003 tubes, and then ran the reactor between 10 and 15 MW on 3 October. Gradually, the reactor was raised through intermediate levels to 38 MW in early October. These experiments revealed the need to increase the loading to achieve design levels of 250 MW and to counter the xenon effect. Through October and November, the same cautious loading, checking, and loading of more tubes continued, finally bringing the reactivity level to 124 MW at the end of November. In order to bring the reactor to full-scale operation, some 400 tubes, which had been built without water fittings, had to be fitted out, and the work proceeded through December, when finally 2,002 of the 2,004 tubes were loaded.

Later reviews of this experience by Du Pont stressed the fact that the company engineers had taken the conservative approach of building excess capacity in the form of tubes that were not originally planned as needed, despite advice from Chicago to the effect that such engineering conservatism only caused delays.[55] The unsuspected degree to which xenon caused poisoning meant the foresight of incorporating extra tubes in the design proved quite valuable.

To deal with the survival of xenon as a poison in the pile for as long as 10 hours after a shutdown, operators worked out several new startup procedures involving more rapid control rod removal. After a series of scrams (as the shutdowns were already being called), readjustments, and further tests, the design level of 250 MW was finally achieved, with 2,002 tubes loaded, on 4 February 1945.

One later surprise scram came as a result of an off-reservation outage of the power supply on 10 March 1945. As veterans of the early Hanford days

Du Pont engineers designed Hanford's B reactor with an excess of fuel-element slots, which successfully combatted xenon poisoning. (Westinghouse Hanford Company)

BUILDING REACTORS AT HANFORD | 41

Table 1. Reactor scrams in early operations at Hanford

	First 4 mos. of operation	No. of scrams
B reactor	Oct. 1944–Jan. 1945	70
D reactor	Dec. 1944–Mar. 1945	21
F reactor	Mar. 1945–Jun. 1945	8

Source: Memoranda for the File, P Department, pt. 2, bk. 2, Hagley Archives, Wilmington, Del., Hagley Accession 1957, box 58.

liked to recall, Hanford was the only nuclear facility in the United States ever to suffer from enemy attack, for the outage was caused by the collision of a Japanese incendiary bomb-carrying balloon with the local power line, which caused a 2-minute cutoff of power.[56]

Startup of F and D reactors went much more smoothly, with a higher number of initial tubes charged and rapid handling of the xenon poisoning and control issues that had been explored in the B startup. F reactor was brought to the design level of 250 MW within its first week and maintained at that level. The lower number of scrams during the first four months of operation of D and F reactors (Table 1) reflected increasing smoothness of the operation.

MED kept production rates and quantities highly classified. However, declassified records indicate that one of the earliest and possibly the first of the discharges of slugs for experimental refining of plutonium came on 18 January 1945.[57] Periodically, the "hot X-metal" cans were discharged in quantities of several tons and sent to the 200 area at Hanford for plutonium refining. Dates for the shutdown and discharge were sometimes set to coincide with planned power shutdowns by the Bonneville Power Authority or with maintenance work such as purging solids from the water system.[58] Over the period from May to August 1945, kilogram amounts of plutonium oxide were sent to Los Alamos, providing barely enough for the device tested on 16 July at Alamogordo and the weapon dropped 8 August 1945 at Nagasaki.

At the request of J. Robert Oppenheimer, the reactors also produced polonium, a radioactive isotope, for use as a neutron source trigger or initiator in the weapons. By 1 May 1945 operators had charged four of the tubes in D reactor with bismuth slugs for polonium production, for a total of 264 slugs.[59] Greenewalt was not at all pleased and went on record with Compton about what he thought the priorities at Hanford should be, writing in the cryptic style that he had developed over the months of secrecy and compartmentalization. Although he understood that "the use of polonium in connection with the construction of the final unit is not only desirable but necessary," he made his objections clear: "The Du Pont Company is most anxious not to complicate its task at Hanford by the addition of any new ventures." Further, he did not want Du Pont to take any responsibility for the refinement of the polonium, leaving that to Los Alamos. Los Alamos did its own refining from the bismuth slugs.[60]

Groves and Col. Kenneth D. Nichols had been aware of Du Pont's reluctance to reserve any space in the Hanford reactors for polonium production, and both Nichols and Oppenheimer went to some trouble to document the need for polonium and to get in writing their request through Groves to Du Pont. As early as 1943, Oppenheimer had anticipated the need, and Nichols had asked Groves to request polonium production from Du Pont in January 1944, when the reactors were just being built. Colonel Nichols pointed out that with all polonium production concentrated at Clinton, the risk of an interruption of supply was real. It was clear from these exchanges that Du Pont managers at Hanford wanted to document any delays in production, especially those which resulted from decisions by others in the weapons complex.[61]

Reactor Cousins and Nuclear Politics

During the war, work went forward on alternate reactor designs at Chicago and Argonne, with various configurations of moderator, coolant, fuel enrichment, and fuel arrangement. In addition, planning groups considered possible future uses for reactors: production of plutonium and a variety of isotopes, testing of physical principles, electrical energy generation, space flight, and aircraft and ship propulsion.

Farrington Daniels developed a concept of a beryllium-moderated, helium-cooled reactor, which remained a notional reactor for years. In later years, the "Daniels pile" was sometimes cited as the first "gas-cooled" reactor, although it had been preceded by X-10, which was gas-cooled in the sense that air is a mixture of gases.[62] The concept of both the Daniels pile and X-10 had been preceded by early thinking about a graphite-moderated, helium-cooled model developed by Moore's Engineering Council. In Canada, the Montreal Laboratory studied the possibility of a heavy-water-moderated and -cooled reactor, and the Argonne group worked fairly closely, within the limits of international agreements, with the Canadian group. By

1945, the Canadians had built a zero-power heavy-water-moderated reactor, and Argonne had CP-3, the first American heavy-water-moderated reactor.[63]

Thus, before the end of the war, the lineal ancestry of several later reactor types had been established. Driving the original work had been the push to build the production reactors, leading to the construction of the exponential piles by Fermi; the operative CP-1 (rebuilt as CP-2); the experimental heavy-water model CP-3; the intermediate X-10 at Clinton, which actually produced small quantities of plutonium; and the massive full-scale production reactors B, D, and F on the Columbia River at Hanford. Du Pont had kept its focus on producing enough plutonium to win the war and not on engaging in a range of research, producing isotopes, or building a position in an emerging industry through the development of a full range of reactor types. Du Pont had not encouraged the work on heavy water, since that would distract from the company's wartime nuclear mission, and it had discouraged the diversion of B, D, and F reactors to polonium production. As a consequence, the United States had developed its richest experience in one type of reactor: water-cooled, graphite-moderated piles for the production of plutonium. The heavy-water alternative remained a strictly experimental operation at Argonne.

The politics that would later surround reactors of all kinds—whether production, propulsion, experimental, or power-generating—were already beginning to show their shape in the restricted and small community of those with the knowledge and the security clearances to recognize some of the implications of their work. Decisions and choices that later became subjects of national debates were explored in the confined community of science and technology policy makers within the Manhattan Engineer District.

Perhaps the first such nuclear political or "technopolitical" issue had been simply how best to utilize atomic energy. The Army and the Army Air Force saw the appropriate use of the energy released from matter as a weapon: a bomb. When Du Pont took on the mission of building the reactors, it held strictly to the agenda of building and operating one type for one purpose: graphite-moderated, water-cooled, plutonium-producing. During the war, the Navy hoped to harness nuclear energy through reactors as a propulsion device for submarines and ships. These efforts centered at the Naval Research Laboratory under Ross Gunn. At Chicago, those looking ahead to the postwar years and concerned with peaceful uses of nuclear physics could see several immediate applications: reactor fission products could serve as substitutes for radium in radiation treatment; other radioactive isotopes produced in reactors could serve as tracers in a wide variety of biological, geologic, and medical research. Furthermore, atomic energy might serve as a source of electrical power through the harnessing of the waste heat of reactors and its conversion to steam. Such choices were still in the future, but scientists had anticipated them for decades.[64]

In the early, partially organized efforts of the Chicago atomic scientists to affect policy lay the roots of the later Federation of American Scientists, known for its *Bulletin of Atomic Scientists*, with its concern that atomic energy be brought under international control and be converted to peaceful uses. At Chicago, the emerging group of nuclear physicists doing engineering work on practical problems dubbed their new field "nucleonics," coupling their practical plans with idealistic visions of a nuclear future. Groves, with his distaste for the independent-minded Szilard and his military objection to scientists who took it upon themselves to question orders and authority, regarded some of the free thinkers at Chicago as being close to disloyalty. By the end of 1945, the built-in fissures in the hastily assembled nuclear community were showing signs of widening as many scientists, frustrated at military control, turned increasingly to political action.[65]

The technology of the first three production reactors was to an extent applied physics, but it was much more. The specific design of the reactors had been shaped out of the interplay between the academic-style committees established at the Met Lab, Greenewalt and his Du Pont engineering staff with their approach based on the nylon-development model, and the demanding schedules set by General Groves. The choice of the conceptual designs; decisions as to location, shielding, cooling, control, and handling of effluent; and the solution of the xenon-poisoning crisis were collective products of the emerging human institutions of the weapons complex culture and the new profession of nuclear engineers.

3 • Contracting Atoms

In the postwar years 1946–49, international events helped convince President Truman that the threat to peace continued and that atomic weapons were crucial to a strong defense. Winston Churchill, long an advocate of a vigilant stand against the expansion of Soviet influence in Eastern Europe, warned in March 1946 that an "Iron Curtain" had descended, as Communist-controlled local governments took power under the aegis of Soviet Army forces in Eastern Europe. The year 1946 saw a minor crisis as Soviet forces delayed their departure from northern Iran. Britain, suffering from the destruction and economic ravages of the war, announced it could no longer provide troops to sustain the pro-Western regime in Greece against a Communist-led insurgency; in response, President Truman announced the Truman Doctrine and obtained congressional approval of funds for military assistance to both Greece and Turkey. Soviet distrust of the Western Allies as they moved toward uniting their three occupation zones of Germany led to the blockade of land routes to Berlin in 1948–49, further heightening tensions. Truman's responses to these and other developments helped shape the nature of the postwar world, confirming the division into two increasingly hostile armed camps.

In the years 1946–49, the United States held a monopoly on the nuclear weapon. That monopoly provided a certain assurance, constituting a backup to the nation's strong stands in Western Europe. Thus, policy required that the nuclear complex be kept in place, maintained, and upgraded to ensure that a stockpile of nuclear weapons would be available. The nuclear weapon would be the central element of the emerging preparedness doctrine or ideology.[1]

The immediate postwar period was one of transition for production reactor management and planning. The issue of civilian versus military control of atomic energy was fought out and resolved by Congress with the establishment of a new civilian agency, the Atomic Energy Commission (AEC). Du Pont, as anticipated in its contract, rapidly departed Hanford, consciously abandoning any possible central role in the newly emerging field of nucleonics. General Electric replaced Du Pont as the operating contractor there.

The transition to a new institutional framework led to delays, false starts, and indecision, in contrast to the quick and decisive pace of Groves and his advisers during the war years.

At the technical level of conversion to peace, the production reactors had to be changed over from wartime crash production to sustained, long-term production to maintain American atomic weapon capacity. Like most of the rapidly erected conventional weapons plants, aircraft factories, and shipyards of World War II, the first reactors, B, D, and F, had not been designed for permanent operation. Temporary facilities, even though situated in the dry eastern Washington climate, soon deteriorated.

Technical modifications to those existing reactors and a series of innovations on two new reactors built in the postwar era reflected the altered conditions of policy and management. The mechanical changes were incremental, not revolutionary, and represented solutions to newly discovered problems and several means of increasing production. The decisions about existing and planned production reactors came out of the context of increasing Cold War tensions, institutional change, some scientific advance, and solutions to problems in safely maintaining and increasing the nation's nuclear stockpile.

Institutionalizing Control of Atomic Energy

When President Truman announced that an atomic bomb had been dropped on Hiroshima, the issue of control of the nuclear weapons complex immediately entered the public forum, leading to political struggles over the exact shape of atomic energy legislation in the United States. Deciding on the form of the new institution and getting it in place took 20 months, from September 1945 through April 1947.

For men like Gen. Leslie Groves and Secretary of War Robert P. Patterson, retaining military controls over the use and development of fissionable materials would ensure that essentially military decisions would be reached in the military sphere. Such leaders supported the May-Johnson atomic energy bill, introduced to the House of Representatives in October 1945 by co-sponsor Andrew Jackson May, chair of the Military Affairs Committee.[2]

Atomic scientists who anticipated dramatic discoveries and potential peaceful uses of nuclear energy hoped the postwar nuclear arrangements would include opportunities to conduct basic research. Furthermore, Congress and the president wanted assurances that the new agency controlling atomic energy would remain accountable to elected representatives, not simply to the military; decisions to manufacture, deploy, and ultimately use the most awesome weapon of all time, in their view, could not be trusted to military officers. The May-Johnson bill, with its restrictions on outside research and its seemingly independent body of military decision-makers, distressed both scientists and advocates of civilian control. In response, Senator Brien McMahon introduced an alternate bill.[3]

The McMahon bill encountered a difficult political environment. Revelations in Canada in February 1946 that a spy ring had relayed atomic secrets to the Soviet Union made continued military control of atomic energy, with its security arrangements, seem a good idea. McMahon succeeded in getting his bill enacted only after agreeing to a military liaison committee, among other compromises. President Truman signed the Atomic Energy Act on 1 August 1946.[4]

Under the act, the president would appoint five commissioners, with one designated as chairman. By October 1946, Truman had chosen his five appointees, with David E. Lilienthal as chairman. Authority could not be immediately transferred from the Manhattan Engineer District to the new civilian agency, which under the act officially opened 1 January 1947. When the new session of Congress convened in January and took up appointment confirmations, Lilienthal and the other commissioners had to serve in an "acting" role until their confirmation in April 1947.

Before his appointment to the AEC, Lilienthal had been director of the Tennessee Valley Authority (TVA). The other four commissioners were Lewis L. Strauss, a partner in Kuhn, Loeb & Company; Sumner T. Pike, a former member of the Securities and Exchange Commission; William W. Waymack, editor of the Des Moines *Register and Tribune* and public director of the Federal Reserve Bank in Chicago; and Robert F. Bacher, a physicist who had worked at Los Alamos during the war. Bacher was the only commission member with a technical background in the nuclear field. The lack of scientific or engineering experience at the topmost management level

In April 1947, Gen. Leslie R. Groves (*left*) turned over control of Oak Ridge and Hanford to David E. Lilienthal, first chair of the newly created Atomic Energy Commission. (U.S. Department of Energy)

of the commission appeared to some scientists to be a basic weakness in the new structure.[5]

In common usage in the period, the term "Atomic Energy Commission" could refer either to the governing group of five commissioners or to the agency as a whole. The Military Liaison Committee (MLC) provided the AEC with its formal access to the military. Composed of six representatives appointed by the Secretaries of War and the Navy, the MLC assisted the AEC first with the transfer of responsibilities from MED and then with issues relating to security, fissionable materials, and research. The MLC provided a channel through which knowledgeable military officers could provide the benefit of their experience and advice; the committee was not intended as a means of providing military control.[6]

The General Advisory Committee (GAC), consisting of nine civilians appointed by the president, was established to serve as a source of advice from experienced nuclear physicists. Most of these physicists had worked on the Manhattan project. In addition to Isidor Rabi, a Nobel Prize winner, the GAC included James Conant, Enrico Fermi, Hood Worthington of Du Pont, and Glenn Seaborg, a chemist credited with the co-discovery of plutonium during the war. The GAC provided the Commission with advice on technical, scientific, and policy matters relating to fissionable materials, production, and research and development. J. Robert Oppenheimer served as the committee's first chairman, and he used the position to attempt to steer nuclear policy as a whole. Originally meeting at 2-month intervals, the GAC developed a close working relationship with the Commission as it provided technically informed policy advice.[7]

In this formal fashion, the different cultures that had been so uncomfortably merged by Groves in the MED survived in the new agency, somewhat reflected through the advisory committees. The military men, with their concerns for secrecy, compartmentalization, strength of forces, and the meeting of definable objectives could at least communicate their viewpoint through the MLC and through the military officers who served as directors of military applications. The civilian scientists and industrial engineers, with their concern for long-range humanitarian issues and their interest in sponsoring theoretical physics research, had a somewhat more influential voice through the GAC. The difference was never clear-cut or absolute, and was usually one of emphasis, for the scientists understood and supported the central weapons mission, while the MLC respected the need for continued research. Ultimate decisions rested with the commissioners. A potential for disagreement between the advisory groups was built in; it later surfaced over the issue of how to proceed with the development of a fusion weapon, the hydrogen bomb, as well as on less spectacular issues, such as reactor siting and construction.

While the MLC and the GAC served in advisory roles to the Commission, the congressional Joint Committee on Atomic Energy (JCAE) represented the first legislated attempt to structure broader public participation while maintaining security restrictions, providing a channel for input from the political side. Although McMahon, as a Democrat, did not serve as the committee chair in the Republican-dominated Eightieth Congress of 1947–49, he was the senior Democrat on the committee and chaired the committee later when Democrats gained control of the Eighty-first Congress in 1949. It was largely through his vigor and commitment that McMahon ensured congressional participation in literally hundreds of policy issues related to nuclear energy. The JCAE gradually increased its authority, establishing the right to consider all atomic energy bills and resolutions introduced in Congress and to hold hearings on all AEC activities. The Commission was required to keep the Joint Committee "fully and currently informed" and to report regularly on the status of its properties, facilities, contracts, personnel, financial dealings, and future plans. The JCAE became a forum where many of the early debates over production reactors were aired; its records, now for the most part declassified, provide a view of the changing and sometimes chaotic policy environment in which the technology of production reactors evolved. As congressmen unfamiliar with the basic physics of the reactors and bombs moved into areas of policy and management, they, like the military men, frequently focused on issues of security, access to information, and budget, rather than on issues of research. In effect, they brought their political skills to bear on a field in which they had few of the skills required to speak to issues of research policy, and made contributions in what they regarded as their areas of special competence.[8]

Transferring the nuclear enterprise to the AEC necessarily resulted in some cultural and institutional continuities. Like the Manhattan District, the AEC had extraordinary powers given to it by the Atomic Energy Act, to produce fissionable material, to man and operate production facilities, and to control the materials produced. Practices and patterns established by Groves transferred to the Commission, including the whole government-owned contractor-operated (GOCO) system, the operational methods for the physical facilities, and the division of responsibilities between sites.

In addition to the formal institutional embodiment of the military, scientific, and political perspectives, there were some less formal aspects of the organization that provided cultural continuity. Many of the individual managers, engineers, and technicians who had designed, built, and operated the first nuclear production facilities during World War II remained at the sites even as their corporate affiliations changed. Faces at headquarters might come and go, but at the practical level of daily work, many of the patterns established during the war lived on. Forty years later, some of the men who had been present at the xenon poisoning of B reactor still worked at Hanford as retired consultants and specialists in declassification work.[9]

In early 1947, the Commission began to organize for further development of nuclear energy. In the process, the AEC started to develop cultural traits or practices that departed in a few significant ways from those established under Groves. One important difference between the new agency and the wartime operation was increased reliance on local authority. Since the Commission had only a small staff, such reliance was to a degree the product of necessity. However, Lilienthal also drew from his own experiences as administrator of TVA; there, he had joined with others in trusting local administration over Washington bureaucrats. Though Washington still attempted to come up with an overall policy, the individual sites developed varying degrees of independent and local cultures, reflecting the approach of the operating contractor, the local mix of science and engineering, and emerging ties to local communities and politicians. In time, as contractors changed, each facility developed a slightly different institutional personality, reflecting its own history of shifting contractors and isolation from the rest of the nation and from the other sites.[10]

Perhaps the most significant difference between the Manhattan District and the new agency was that the AEC did not have a General Groves, providing at once the drive, the control, and an understanding of the engineering tasks, all concentrated in one person. Rather, expertise and authority was somewhat more diffuse. The new commission appointed Caroll Wilson as general manager. Wilson had little direct experience with atomic energy but considerable background and contacts in the administrative side of the emerging network of government-funded scientific work. He had served as an adviser and assistant under Karl Compton, president of the Massachusetts Institute of Technology, and under Vannevar Bush when Bush had been dean of engineering there. He had followed Bush to the National Defense Research Committee and the OSRD.[11] The Commission had several other leading personalities with high technical competence. Commissioner Robert F. Bacher, a nuclear physicist, had worked as a division director under Oppenheimer at Los Alamos. Through the GAC, the Commission had access to a group of renowned physicists with pertinent experience.

The significant change was the transfer to collective management of a far-flung industrial network, with neither the urgency of war nor the personal leadership of a single personality. The huge technical complex that the AEC inherited from MED had already begun to develop local autonomy even before the war ended. Under Groves, decisions on both large and small matters cascaded from his office in the form of orders and angry demands for responsible performance. After the establishment of the AEC, with its oversight and review from the congressional committee and its outside military liaison and scientific advisory committees, it was sometimes difficult to obtain quickly a single firm technical decision among options. As a consequence, technological choices as to new reactor construction or reactor utilization were made, rescinded, discussed, made again, and then discussed again in several forums. Technical people in the field, always suspicious of

Washington, had good reason to be impatient and frustrated. And likewise, the AEC and the JCAE had reason to grow frustrated with the sometimes maverick units in the vast weapons complex and with the process of establishing clear lines of communication and direction with the operating contractors.[12]

General Electric Replaces Du Pont at Hanford

During the same period that Congress established the AEC, the operation of Hanford changed hands. Du Pont's contract with the War Department specifically granted the company the option to leave Hanford 9 months after the cessation of hostilities. Du Pont agreed to extend its participation until 31 October 1946 because of the delays in setting up the Atomic Energy Commission.[13]

Though Groves had known Du Pont's intention to depart from Hanford from the beginning, he still tried to convince company president Walter Carpenter to rethink or delay the decision. Appealing once again to the demands of "the national welfare and the national defense," Groves applauded Du Pont on the "wealth of experience" it had acquired from the project. Secretary of War Robert Patterson echoed Groves's plea, stating that the loss of Du Pont would result in "great material loss" both to the project and to the country. But Carpenter held fast. Carpenter and other Du Pont executives believed that it would be years before a civilian market developed in the nuclear field and that it would be difficult to recruit the nuclear physicists, who remained suspicious of Du Pont and corporate activity, in any case. In effect, Du Pont sought to turn its research energies to more promising new products, those with greater potential markets.[14]

Groves considered several companies as a replacement for Du Pont. Monsanto had some experience with atomic energy, since it had worked as operating contractor at Oak Ridge during the war, but it was a small company with few qualified personnel. Its limited resources were already taxed with its operation of Clinton Laboratories and its developmental work on two experimental pilot units.[15]

Groves considered General Electric a better choice, an "outstanding American company" with interests in the future applications of atomic energy. But General Electric was not easily persuaded by Groves's overtures. The uncertainty about the future of the entire nuclear weapons program in early 1946, as Congress debated the establishment of the Atomic Energy Commission; GE's own plans for reconversion from wartime to peacetime production; and corporate concerns about liabilities—all kept the company from immediately accepting the Hanford task.[16]

Continued prodding from the War Department eventually convinced General Electric president Charles Wilson. On 28 May 1946 Wilson finally agreed, concluding that it was of "tremendous importance" to the national interests that the country maintain "preeminence" in atomic energy, both for military and peaceful uses. Wilson qualified his acceptance with two stipulations. He required that the contract contain a provision freeing the company of its obligations in case the atomic energy legislation imposed conditions that in GE's "sole judgment" the company considered "unacceptable." In addition, General Electric expected full recovery of all costs incurred in connection with the contract and protection against any liabilities, since hazards of "an unusual and unpredictable nature" were involved. Clearly, Wilson respected the risks involved in operating the plutonium production reactors at Hanford and sought protection similar to that extended to Du Pont during the war. Patriotism did not mean that a multi-billion-dollar corporation would risk its existence.[17]

Transfer of responsibilities from Du Pont to General Electric proceeded without major difficulties, and Du Pont formally withdrew from Hanford by 1 September 1946. Groves once again formally thanked Du Pont for its contributions, which "resulted directly in the saving of many thousands even tens of thousands of American lives." With that statement, he reflected the position which he, Truman, and most Americans adopted in regard to the morality of the decision to drop the weapon on Japan. Groves singled out for particular praise Crawford Greenewalt, who, in Groves's view, had succeeded in translating "meager scientific data" into the information upon which the Hanford production facilities were designed.[18]

With Du Pont gone from Hanford, General Electric initially contracted with the War Department until the commissioners were confirmed and could legally approve the new arrangements. With the interim General Electric contract due to expire on 30 September, Groves had to extend the terms twice—first to 30 November, and again to 30 January 1947—to allow the still-unconfirmed acting commissioners time to study the contract fully. According to the contract, GE would operate Hanford, conduct research and development on process operations, and design and construct additions to the site. The contract also instructed the company to establish a research laboratory, later called Knolls Atomic Laboratory, in Schenectady, New York, which was part of the inducement offered by the Commission to

General Electric. At one level or another, GE stayed on at Hanford until 1968.[19]

Though GE eventually mastered the technical difficulties involved in the operation of the first three production reactors, it did so without much central policy direction from the AEC itself at first. The AEC was slow to take hold of its responsibilities even after confirmation of the Commissioners in April 1947. Congressional leaders in the newly created joint committee also needed time to familiarize themselves with the intricate issues of nuclear weapon production; some never succeeded in grasping the basic physics involved. Nor did GE have the impetus or urgency of the war to justify new research and design changes. Under such conditions, both the planning and the management of Hanford operations reflected a sort of disjointed, on-again, off-again progress, and a variety of local, immediate decisions, rather than the clear sense of overall mission that had driven Groves, Greenewalt, and Du Pont. In the face of a national desire for demobilization, the lack of clear mission was to be expected. The company's first priorities, while waiting for decisions from the Commission, were to keep the piles operating, to deal with deferred maintenance, and to make some moves toward improving the living and working conditions in the hastily built facility.

Keeping the Piles Running

By early 1946, the original three piles, which had produced barely enough plutonium and polonium for war uses between September 1944 and August 1945, were showing ominous signs of wear and tear. Under sustained operation, the graphite core of each reactor had expanded and consequently had begun to distort the aluminum tubes that contained the uranium slugs and through which the cooling water flowed. The first three reactors had produced enough plutonium for two weapons in about six months of full operation. Production of enough plutonium for weapons tests and to build even a modest stockpile required continued operation, yet the swelling of the reactors suggested they might soon reach the end of their useful lives. In response to these worsening conditions, the Army put B reactor on standby on 19 March 1946 and reduced the power on the other two piles, D and F, in an effort to conserve their lives.[20]

Putting B on standby served another backup function. Given that polonium, crucial as a part of the initiator in early bombs, has only a 138-day half-life, closure of all of the reactors at once for a period of a year or more for rebuilding could easily have led to the elimination of the United States as a nuclear-armed power, at least until a new reactor could be put in operation to ensure a steady supply of material for the initiators. Without continued reactor operation, polonium decay would render any existing weapons in the stockpile useless in only a few months. Thus there was a need to hold a reactor available in reserve until more reactors could be built.[21]

Shaping Plans

With B reactor out of production and the other two piles running at decreased power levels, plutonium manufacture fell off sharply. Facing this situation when it first convened, the Atomic Energy Commission came to grips with its first issue in production reactor planning and management. The Commission determined that work at Hanford should concentrate on three major objectives: (1) prolonging the useful life of existing equipment through rehabilitation and efficient operation; (2) building replacement piles and additional facilities for increasing production; and (3) developing new and more efficient techniques for operating the piles and processing their products. Lilienthal, in setting a "long-term Commission agenda" early in 1947, noted that operating five piles at Hanford, each the size of the original three, was conceivable. In response, General Electric developed a plan for addressing these goals, including a research program to study the radiation stability of graphite and a construction program.[22]

The weapons stockpile had deteriorated to nothing. When Lilienthal met with President Truman in March 1947 to review the status of the weapons complex, Truman was shocked to discover the closely guarded secret that there were no operable weapons at all! A later count established that there were 13 atomic weapons in the arsenal by the end of 1947; by 1948, the number climbed to 50. By the standards of only a decade later, these numbers hardly constituted a "stockpile," but rather, represented a small collection of hand-crafted devices.[23]

At the beginning of 1947, AEC commissioners and the members of the GAC had doubts about GE's supervision of Hanford. Walter J. Williams, who had become director of all MED production operations under Groves and who served as production director for the new commission, filed reports critical of GE performance. Devoting most of his time to field assignments, including Hanford, Williams reported with disapproval that General Electric was first concentrating its construction efforts on building more permanent housing units and storage tanks for radioactive waste, instead of focusing on the production and separation facilities to meet the Commission's agenda. The GAC, reflecting continued academic-scientific and military skepticism about the efficacy of industrial management, noted in its second meeting "some doubt" as to whether General Electric could handle building replacement piles at the Washington site. Tensions between headquarters and the apparently unresponsive corporation mounted rapidly in the first months.[24]

While Williams and other staff and commissioners placed a high priority on construction of reactors to replace the aging B, D, and F piles, they also concluded that the chemical separations processes used at Hanford during the war needed a major overhaul. These treatment plants recovered plutonium, but did not save the unconverted remaining U-238 in the slugs. In 1947–48, the AEC grew concerned that available uranium ore supplies were not sufficient and did not think it wise to regard the unconverted U-238 as a waste product to be dumped. At the time, the radioactive hazard of the wastes, stored in underground steel tanks, was of less concern than the possibility of shortages of uranium ore. The AEC wanted Hanford to adopt a chemicals separations process called redox, which could recover both plutonium and uranium-238, rather than only plutonium. Early in 1947 the GAC recommended that first priority be given to redox, with "pile construction nearly the same," reflecting the perceived shortage of uranium ore.[25]

In an early example of somewhat divergent advice from the GAC and the MLC, it was the military side, surprisingly enough, that recommended a go-slow policy on production facility development. In an April 1947 Commission meeting with the MLC, Groves expressed his dissatisfaction with the progress of AEC management. He rejected the idea that new piles had to be built immediately. First, he wanted the agency to undertake a complete survey of the raw materials situation, since the natural ores existed in such limited quantities. In addition, he believed that a large reactor construction program needed careful planning and assessment by the individuals conducting the work. Groves, like the GAC, was not confident that General Electric and its subcontractors had the necessary competence, especially so soon after taking over the Hanford reservation. Finally, Groves wondered if so many atomic bombs, and the nuclear material fueling them, were really necessary from a military point of view. In light of the time it would take to actually build another pile, Groves advised the Commission to exercise restraint and wait for the first pile to fail. Despite his reservations, the Commission decided to proceed "vigorously" with plans for new reactors.[26]

In the face of criticisms, GE vice president Harry A. Winne defended the company's performance before an executive session of the Joint Committee on Atomic Energy in June 1947. He pointed out the magnitude of the undertaking the Commission expected of his company. Building at least two replacement piles, as the Commission currently planned, as well as the redox plant, required large amounts of steel, cement, and other materials. Without a war emergency, the company and the Commission would need to induce industry to supply these quantities. Winne also noted that the company that had produced the pure graphite during the war had since closed its graphite operations, making it necessary for the Commission to urge a restored capacity or find another supplier. Winne commented that redox still needed further research to ensure large-scale application.[27]

Problems and Solutions

In the meantime, the only piles in operation, D and F, continued to bulge from the top and sides due to continued graphite expansion. During the war Eugene Wigner at Chicago had identified expansion in graphite as a possible effect of intense heavy-particle radiation. He noted in the Met Lab's December 1942 monthly report that a neutron produced in the fission process possessed enough energy to displace a carbon atom, which, in turn, used its gained energy to dislodge about 2,000 surrounding atoms. Called the Wigner effect, this displacement of atoms from their equilibrium position in the crystal lattice by momentum transfer left vacancies in the benzene structure that characterizes graphite blocks. Interstitial atoms filled the gaps left by the displaced carbon atoms, leading to an expansion of the crystal and an overall increase in size. Wigner and his fellow Met Lab scientists knew that the graphite blocks would exhibit effects from this expansion, but they were uncertain of its specific manifestations or its rate. By 1946–47, with the warping of the aluminum process tubes, General Electric engineers came to see the development of the Wigner effect in the piles.[28]

Ejection and injection of slugs became increasingly difficult because of the bent aluminum tubes from the graphite swelling. General Electric considered replacing the tubes, but this procedure would involve shutting down the reactors for several months and exposing workers to the danger of radioactivity as the interiors of the piles were opened for reconstruction. Yet, the bending of the control rod tubes represented an extremely risky situation if it reached the point of delaying control rod insertion because the rods were essential to shutting down a reactor in an emergency as well as for day-to-day operation. Without a solution to the problem of swelling, the water-carrying process tubes would eventually break, damaging the piles and probably putting them out of service permanently. Winne estimated that the piles would not last longer than two to five years. A solution was needed, and soon, if the United States was to retain its standing as the world's first and only nuclear power.[29]

The Commission and its advisors considered building new reactors as an answer to the problem. The GAC suggested building two completely new reactor areas at Hanford, but such a plan would require more time and labor than constructing replacement units close to the original piles. Nevertheless, locating replacement reactors near the old ones, in the eyes of the Military Liaison Committee, presented its own special difficulties. The planned replacements, which would share waterworks with the existing reactors, could run only if the originals actually failed, since the waterworks could supply only one operating reactor at a time. In addition, the proximity of the planned replacement reactors to existing reactors increased the risk of operating accidents and the vulnerability to enemy attack. When General Electric discovered that B reactor, which had been placed on standby to extend its lifetime, actually was deteriorating faster than the operating piles as a result of rusting parts, the Commission tentatively decided in October 1947 to build three replacement units and two new production reactors.[30]

Meanwhile, through 1947, General Electric researchers found that annealing the graphite provided at least a temporary solution to the worsening problem of expansion. At higher temperatures, the interstitial carbon atoms displaced by the neutron burst had shorter lifetimes, allowing them to slip back into the crystal lattice more quickly. If the reactors were run at 570°F and then slowly cooled, the displaced atoms diffused and found vacancies to occupy within the crystal structure. The irradiated graphite recovered its proper structure and stopped growing.[31]

GE's success in addressing graphite swelling through annealing convinced the AEC that the original piles would not fail suddenly but rather gradually, if they failed at all. As a result, the commissioners scaled back their tentative construction program, authorizing in December 1947 one replacement pile at D (to be called DR) and one at a new pile area, named H. This policy decision reflected a mix of military, mechanical, scientific, and engineering considerations that set a pattern for many of Commission's actions with regard to production reactors.[32]

As the Commission determined the scope of its construction programs at Hanford, it also recognized that the site needed a local federal manager to ensure that the AEC's decisions were implemented by the contractor. Despite GE's successes in solving the graphite swelling, the Commission remained unconvinced that the company would be responsive in building replacement and new reactors promptly. Carroll Wilson, the AEC's general manager, suggested the appointment of Carleton Shugg as local federal representative. Shugg was a vice president of the Todd Shipyard Corporation and had the energy and skills to see building projects to completion within a limited time. He arrived at Hanford on Labor Day 1947, and within 7 months, the site was showing signs of his impact. He had DR's main building going up and H site clearance under way. More than ten thousand construction workers thronged to Hanford, a stark contrast to the quiet days immediately following the completion of the original three piles and to the confused lack of direction through 1946 and early 1947. Shugg's success led to his appointment as deputy general manager in Washington, and his replacement, Frederick Schlemmer from TVA, followed through on Shugg's start. Policy decisions could only be put into motion with the right personnel.[33]

Early in 1948, tensions in Europe escalated. On 1 April, the Soviet Union denied land access to Berlin to the Western occupying powers in Germany. During the Berlin blockade and the American-led airlift of supplies to the city, which continued through 30 September 1949, American military leaders wanted a guaranteed supply of atomic weapons. In light of GE's improvements at Hanford and the Joint Chiefs' requirements for weapons materials, the AEC authorized reactivation of B reactor by July 1948. By the end of the year, the GAC heard from Williams that "things were now getting into line" at Hanford, a recognition of the company's efforts to prolong the lifetimes of the original piles. In fact, actual production levels exceeded scheduled amounts.[34]

As the Commission managers worked to get construction under way, General Electric based its plans for DR and H on the blueprints used by Du Pont for the original three reactors, but with several small variations. Construction of DR had top priority, though scheduling often dovetailed with H in order to facilitate procurement of critical supplies. Experience gained from building DR was then applied to H. With an eye toward heightening safety, especially in recognition of the fact that General Electric expected to run the reactor at higher power levels than the original design called for, the numbers of horizontal control rods and vertical safety rods were increased. For both DR and H, each graphite brick was machined with slightly concave faces, compensating in part for the graphite expansion experienced in the other piles. Construction proceeded rapidly, and the Commissioners soon faced the question of deciding when to start operating the new reactors.[35]

By March 1949, General Electric's rapid construction of DR and H and simultaneous resolution of the graphite expansion problem in the original pile at D had created an ironic dilemma. The DR reactor, originally intended to replace the imminently failing D, was now almost fully built but without a separate waterworks system to allow its startup. In an effort to test the feasibility of running D and DR simultaneously, GE had increased D's waterworks to handle 40,000 gallons per minute. However, both reactors running at full capacity required a total of about 64,000 gallons of water per minute. If the Commission wanted both reactors operating simultaneously at full blast, it had to address the need for a new waterworks for DR.[36]

Carroll Wilson, speaking before the joint congressional committee's executive committee in March 1949, suggested the possibility that F reactor's waterworks could be made available for DR. When this reactor was built during the war, its central region had been loaded with some of the highest-density graphite used in all of the piles. This same area, according to Wilson, was currently experiencing the greatest amount of expansion, and GE was not certain that the annealing measures that had been used successfully to solve graphite expansion in B and D would leave F fully functional. If not, then F's water could be pumped across the desert to DR, in effect making DR a replacement for F, not D.[37]

Another possibility, of course, was to make DR a separate site with its own waterworks. However, the Commission and the JCAE were unwilling to ask Congress for the necessary funding until production requirements forced their hands. As the reactor picture stood in 1949, D was "perking along satisfactorily," B and F continued in operation, DR remained unloaded and in standby, and H reactor was slated for full operation once it was finished.[38]

More Production through Power Upgrades

One promising way for increasing production while awaiting completion of H reactor and a solution to the DR waterworks problem involved running the older piles at a higher neutron flux and thus at a higher temperature. Higher power ratings would lead to increased production and, because the higher power levels also reduced swelling, possibly to safer operation. Yet running the reactors at once hotter and at a higher flux did not, intuitively, seem safer, generating concerns among the GAC scientific advisors who studied the issue. B, D, and F were originally designed to run only at 250 MW. Significant departures from this rating needed study and controlled observation in order to ensure safety.

In order to promote higher graphite temperature, General Electric operators experimented with varying concentrations of carbon dioxide–helium gas. Originally, pure helium had been used in the piles to keep the graphite from catching fire from the oxygen in the air when run at high temperatures. With further experimentation, though, Hanford operators found that CO_2 exhibited the same inert qualities as helium but with poorer heat transfer, allowing greater control of graphite temperature. Therefore, a mixture of helium and CO_2 was used. Using annealing with other modifications, it soon became possible to operate B, D, and F reactors at much higher power levels than their original design levels.[39]

While increased temperatures kept the graphite from swelling further and, in some cases, even shrank it to its original size, hotter conditions limited the already short time periods in which fuel slugs could remain in the piles. A standard fuel element consisted of a solid rod-shaped piece of uranium 1 inch in diameter and 8 inches long that was soldered into an aluminum can. Under higher heat, the contained uranium caused blistering and sometimes rupturing of cans on a fairly frequent basis. By 1947, operators at Hanford developed a series of methods for detecting slug swelling and failure. One of these was a simple optical test: the decrease in light intensity through the process tube indicated the presence of blisters on the cans. Another method involved measuring the level of radioactivity discharged from the process tube water. An alarm sounded when the reader encountered a sudden increase in the number of neutrons, indicating that a slug had ruptured and released some its radiation to the cooling water. The panelit gauge allowed immediate tracing to the particular offending tube and slug.[40]

In the late 1940s and early 1950s, researchers at the Massachusetts Institute of Technology, Battelle, Argonne, Schenectady, and Hanford investigated slug failures, looking specifically at how uranium acted in fission reactions. The results of these experiments provided important information for improving the designs of the fuel elements, which could then not only tolerate the greater heat but also produce more plutonium. Over the next years, several relatively simple mechanical expedients adapted the reactors to higher operating temperatures and higher power levels. Without any spectacular invention, but through dozens of minor incremental innovations, the Hanford reactors evolved under the hands of the GE managers and the continuing cadre of engineers.[41]

The GAC had recommended the creation of a safety panel, composed of a group of "disinterested experts," in June 1947 when it had grappled with the difficult problem of evaluating potential dangers in reactor operations. The panel thus formed, Reactor Safeguard Committee (RSC), closely examined the consequences of each of the "gradual stepwise increments" in power levels. This separate advisory body, headed by Edward Teller, studied reports from operating personnel.[42]

Striking a satisfactory balance between the AEC's production demands and the RSC's need to ensure adequate safety proved a sensitive issue when the power upgrade program started. In October 1948, the GAC expressed concern that Teller's committee, barely a year old, might already be acting as a "retarding influence" on reactor development because of its emphasis on "special hazards" as opposed to adequately estimating their "probability" of occurrence. The RSC focused its analysis on how much radioactive material would be released in a single, definable catastrophe and then determined how to limit that release to allowable tolerances by studying meteorological, hydrological, and topographical factors. At the time, there was no generally accepted means of estimating probabilities of reactor failure.[43]

In order to make decisions on power upgrades, the RSC depended for its information on GE. But in Teller's opinion at least, the company had been negligent in providing "specific figures" and other data. In 1949, Teller stated that the committee could not object to proposals for further power increases, since it could not fully evaluate the effects on safety. He warned that the RSC could not share the responsibility for any new operational plans without further consultation. In this tense exchange in early 1950, the RSC formally requested General Electric to forward multiple copies of all reports "having a bearing on safe operation" of the reactors to the Hanford Operations Office for distribution to the committee members.

The early clashes with the outspoken Dr. Teller over safety through classified memoranda bear a surface resemblance to later, more public controversies over reactor safety issues. Yet Teller and his committee supported the weapons mission and the goal of increased production; the RSC's goal was to *ensure* continued weapons production and reactor operation, not to resist production in the name of safety. The tensions between the Teller-led RSC and GE arose from issues of prompt and complete communication rather than from a deeper disagreement over the goal of increasing production. The later public complaints about safety often seemed motivated by opposition to the weapons program itself, with the safety issue used as a means of expressing opposition indirectly. Edward Teller was never accused of having qualms about the value of nuclear weaponry.[44]

By March 1950, the Reactor Safeguard Committee had approved operation of the piles at 305 MW (up from 250) and, in light of encouraging graphite studies, favorably considered an incremental increase to 330 MW. With each power upgrade, the RSC also considered the effect of discharged effluent on radioactivity levels in the river. The committee noted that the allowable increase was still within the tolerance limits for human consumption of water, though the river above and below Hanford had to be assessed to properly measure the effects.[45]

Though graphite swelling, radiation exposure, and general fears of catastrophic incidents necessarily limited how far the operating contractor and the RSC wanted to push the operating levels of the Hanford piles, the source of that power—fuel slugs—needed constant redevelopment in order to withstand the intense heat to which they were routinely subjected. By 1950, adopting results from detailed metallurgical studies of uranium, GE substantially increased exposure times for slugs, up to three times the level feasible in 1946, without encountering blistering or warping. One method, involving the use of 2% zirconium alloyed to the slug, helped stabilize the slug, even under very high exposures.[46] Improved inspection techniques, new water treatment processing, and enhanced instrumentation also extended fuel slug lifetime and increased overall production levels. The technical workers materially reduced slug corrosion by changing the chemical treatment of the pile input water.

Another problem was "boiling disease," which referred to accidental raising of the water temperature in particular fuel-loaded process channels to 212°F. A resulting pocket of steam presented problems when an alternating-phase steam-water system developed in the tubes and caused increased resistance, decreased cooling, and increased slug ruptures. The thin aluminum-zirconium coating on the slugs would expand, blister, crack, and spill the contents into the tubes. Ruptured slugs required shutting down the reactor and then recovering the damaged slug, which might be stuck in a tube. Dealing with this problem led GE to design new control instruments. At the onset of boiling disease, these devices scrammed the piles, reducing the chance of greater potential damage. Local operators also developed an array of special tools for recovering the ruptured slugs, tools which remained on proud display at B reactor 40 years later.[47]

Each of these steps for incremental increases of reactor power levels was significant not just from a production standpoint but also, and perhaps more importantly, from the vantage of achieving greater economic efficiency. By the early 1950s, atomic energy policymakers came to realize that plutonium was costing the United States a great deal of money. Though high capital expenditures for secondary facilities contributed to the overall figure, the AEC supported increased power levels in part because these tended to lower plutonium production costs per gram. G. R. Prout, a General Electric vice president and chairman of the company's nucleonics department, told the JCAE in June 1950 that improved slug designs had allowed higher power levels and had cut raw material requirements by 50%. Not only did the procedures allow for better use of scarce uranium, but the steps led to dollar savings, with the company producing 40% more plutonium per dollar of operating cost in 1949 than in 1947. Although the company operated on a cost-reimbursement contract, its ability to bring cost analysis and procedural improvements to bear began to win it warm support in Congress.[48]

As production reactor management decisions implemented a general policy to maintain the nuclear monopoly, those decisions continued to reflect a complex weave of local technical solutions, scientific findings, economic considerations at the congressional level, engineering options and plans, and questions of personnel. Clearly defining the whole process was the overall place of production reactors in the nuclear arsenal and the perceived need to use that arsenal as a instrument to control Soviet ambitions.

Through the early postwar years, the United States thought it had a monopoly on the winning weapon. Most concerned policy officials believed in mid-1949 that the monopoly had been maintained through a combination of secrecy about the crucial elements of the weapon, limited supplies of uranium, and the difficulty of the weapons-making process. Thus, decisions to keep the reactors operating were taken against a background of an assumed solid nuclear lead over the Soviet Union.[49]

Groves himself estimated at the end of World War II that the Soviets lagged behind the United States in progress toward a nuclear weapon by as much as 20 years. This estimate was based in part on outdated geological maps of the Soviet Union that showed few uranium deposits. The Czechoslovak coup d'état of 25 February 1948, which converted that nation into a Soviet satellite state, appeared ominous; the Joachimstahl uranium mines there might aid the Soviet effort to become a nuclear power. The Berlin blockade, beginning in April 1948, contributed to military concerns about maintaining a weapons stockpile. But events in 1949 and 1950 would completely destroy Americans' assumptions about their lead over the Soviets in the nuclear field.

4 • Flexible Design at Savannah River

The Soviet Union exploded its first atomic weapon in late August 1949 but made no public announcement of the event. In the United States, the Atomic Energy Commission's program of sampling rainwater confirmed the Soviet test by revealing weapons-test fallout. Truman and the commissioners were shocked, for they believed the nuclear monopoly secure. In order to conceal the existence of the monitoring method, Truman waited until 23 September to inform both the public and his cabinet of the news. Truman still appeared not entirely convinced the Soviets actually had a bomb, referring in his speech to an "atomic explosion."[1]

Lilienthal saw some good in the crisis; he hoped "the old spirit of emergency" would be restored, allowing vigorous pursuit of new construction and new scientific advances.[2] Intelligence sources had supported the president's and the nation's belief that nuclear capability in the USSR was still months or even years away. As late as July 1949, the Central Intelligence Agency had estimated that the Russians would not have a bomb until the summer of 1950, and more likely the summer of 1953. Feeling misled by the nation's new espionage coordinating agency, Senator Eugene Milliken of the JCAE reminded his colleagues in October 1949 of the perhaps "innocent" assurances of CIA director Adm. Roscoe Hillenkoeter, less than two months before the explosion, that the CIA "possessed much factual data" about the Soviet's slow progress toward a nuclear weapon. The CIA drolly responded that it was reviewing the data "in light of this development" with an eye toward revising its estimates on the date of Soviet weapons production.[3]

"Little Joe," as American journalists dubbed the Russian device, led the Joint Chiefs of Staff to set new minimum requirements for the atomic stockpile with a demand for increased production. In response, Lilienthal informed Hanford's manager Fred Schlemmer that construction of DR waterworks, which had been placed on hold, should proceed immediately. Lilienthal also stressed the importance of completing and beginning operations of redox.[4]

However, as atomic energy policymakers considered building new production piles in response to the Soviet detonation, they did not immediately support locating them at Hanford. As Carleton Shugg made clear in an October 1949 JCAE executive session meeting, any new reactor construction would throw the balance at Hanford "badly out of whack," since support facilities would have to be upgraded to handle the increased plutonium production. Instead, the committee considered locating new piles at a site other than Hanford.[5]

At the same time the commissioners worried that the current weapons program would be insufficient to address the new Soviet threat. Commissioner Lewis Strauss wrote to his colleagues urging that they take a "quantum jump" in planning by intensifying efforts toward developing a thermonuclear weapon.[6]

The rest of the Commission was less inclined than Strauss to immediately adopt this position. The thermonuclear bomb, or the "Super," had been investigated during the MED period. Scientists, politicians, and even the general public had heard of H-bombs. Even though such weapons had not yet been designed, it was assumed that when developed, one would produce enough destructive power to obliterate an area of 100 square miles. For Chairman Lilienthal and many others, a decision involving development of a weapon 100 to 1,000 times more powerful than the one dropped on Hiroshima required not just consideration of such routine factors as costs, feasibility, and efficient use of fissionable materials, but also "psychological imponderables" and moral issues. As Lilienthal noted, "I regard the matter not as one for the Commission merely, or chiefly, but essentially a question of foreign policy for [Secretary of State Dean] Acheson and the President." The decision would be tough.[7]

In early November, Lilienthal laid out for Truman the thermonuclear situation as the commissioners, individually and as a group, saw it. Following a subcommittee trip to Los Alamos and Berkeley, the Commission concluded that, with a minimum of 3 years' development, "there is a better than even chance it can be made to work." In addition, Lilienthal acknowledged that the Soviets were already familiar with the ideas and would probably complete their development work within a short time frame if they started on it.[8]

Nagging at Lilienthal, however, was the realization that if the United States pursued development of this weapon, it would "intensify in a new way" the U.S.-USSR arms race, ultimately calling into question America's commitment to peace. Lilienthal wanted to keep the country strong, but he was not convinced that hydrogen bombs secured any additional strength over the current atomic weapon stockpile. Instead, he believed that U.S. adoption of the Super would signal to the world that "we have abandoned our program for peace and are resigned to war." And this war, in Lilienthal's eyes, would rely almost entirely on new mass destruction weapons. "A costly cycle of misconception and illusion" would irretrievably focus America's national policy on superbombs as the chief means for protection. Clearly, more was at stake than simply developing a new kind of bomb.[9]

Siding with Lilienthal were J. Robert Oppenheimer and some of his fellow scientists on the General Advisory Committee. The GAC considered the Super completely different from an atomic bomb. Any decision to use such a weapon, the GAC reported, would be "a decision to slaughter a vast number of civilians." In addition, members expressed alarm regarding the possible global implications of releasing radioactivity that would render large areas uninhabitable long after a war. Enrico Fermi and I. I. Rabi told the commissioners that the use of hydrogen bombs would place the United States in a "bad moral position" relative to other countries. For the GAC, thermonuclear devices represented an entirely new and unwelcome stage in the quest for national security. The opposition of the GAC was strong and strongly stated.[10]

On the other side, Lewis Strauss was joined by a few scientists and many politicians who voiced their reasons for supporting the design and construction of thermonuclear bombs. Ernest O. Lawrence, director of the Berkeley Radiation Laboratory, noted that evidence already suggested the Soviets were "well on their way to production," and the United States had "no time to lose." JCAE chairman Brien McMahon wrote to the president that "the profundity of the atomic crisis which has now overtaken us cannot . . . be exaggerated." With the "wholly new order of destructive magnitude" available in the Super, McMahon argued that the military advantage it presented should not be underestimated. Along with severely reducing an opponent's ability to retaliate, the hydrogen bomb promised the psychological benefit of shocking and demoralizing the enemy. In the end, for McMahon, it was a choice between "catastrophe" if the Soviets developed the bomb first and "a chance of saving ourselves" if the United States succeeded before the USSR.[11] Lilienthal, mocking McMahon's emphasis on worst-case scenarios, characterized the senator's views in his diary as "blow them up off the face of the earth, quick, before they do the same to us—and we haven't much time."[12]

Debate within the confines of the AEC and the JCAE soon spread when the *Washington Post* published an article about American development of a "super bomb," which spurred front-page stories worldwide. President Truman, upset that the issue had entered a truly public forum, asked for immediate recommendations from the secretary of state, the secretary of defense, and the AEC chairman. By early January 1950, both secretaries had concluded that the Super had to be built. In addition, the Joint Chiefs of Staff pushed for the weapon as a deterrent force, while the JCAE, marshaled around Senator McMahon, also supported its development. Lilienthal had the support of Commissioners Smyth and Sumner Pike within the Commission itself, although when he appeared at the National Security Council, Lilienthal found himself a lone dissenter from the enthusiasm for the new bomb. Lilienthal argued to the end that the super bomb meant a "headlong rush to a war of mass destruction weapons," but Truman would not hear it. On 31 January 1950, the president signed his advisors' recommendation for developing thermonuclear weapons, saying, "We have no other course." The next generation of nuclear weapons was to be built.[13]

Supplying the Super with Tritium

On the day of the president's decision, the JCAE discussed what adjustments to make to the current production program. A prime consideration was increasing the production of tritium, a radioactive isotope of hydrogen, which had already been produced in small amounts in the Hanford production piles.

Preliminary designs for the new hydrogen bomb called for quantities of tritium to be fused with deuterium (heavy water) for the energy release. The first hydrogen-fusion device tested by the United States—the "Mike" shot in the Ivy series on 31 October 1952—was a cumbersome "wet" device, with supercooled liquid tritium and deuterium in a building-sized refrigeration unit at the Enewetak test site in the Marshall Islands. More than a year earlier, in May 1951, tritium had been employed to test the principle of "boosting" in Operation Greenhouse. In that series, tritium was first used in a plutonium-fueled atomic weapon to make the fission reaction more complete, or to boost the effect.

Hydrogen bombs as later developed consisted of two compartments—the primary, in which a fission reaction took place, and the secondary, where fusion occurred. In the secondary compartment of the hydrogen weapon, solid thermonuclear fuels, which did not require refrigeration, such as lithium deuteride, reacted with the high-energy neutrons produced from the fission reaction and produced more tritium. This bomb-produced tritium in turn fused with deuterium and produced a high-energy neutron, making the most significant energy component of the bomb. Additional fission reactions occurred when a surrounding case of uranium reacted with the high-energy neutrons created from the fusion reaction. The tritium in the secondary compartment was produced from the reaction in the weapon; it did not have to be manufactured in advance in reactors. With the later hydrogen weapon designs, production reactors needed to supply only enough tritium to boost the primary part, the plutonium atom bomb at the heart of the hydrogen bomb. But in the early 1950s, as the weapon design choices were being thought out, first estimates suggested a vastly increased need for tritium.[14]

Tritium production in nuclear reactors required that target slugs loaded with lithium deuteride be inserted into a pile along with fuel rods containing highly enriched uranium (HEU), uranium with a higher-than-natural ratio of U-235. This arrangement was different from that used for producing plutonium, which required only fuel rods using natural uranium, since the U-238 in the rods served as the "target." Furthermore, efficient plutonium production required natural uranium with a high proportion of U-238. Thus, a choice had to be made between efficient reactor loadings for plutonium production (natural uranium) or efficient reactor loadings for tritium production with a lower proportion of U-238 and a higher proportion of U-235—that is, HEU. One or more reactors would have to be set aside and converted to tritium production with the lithium deuteride targets, but such a decision would reduce plutonium production. Hanford's DR reactor, whose waterworks was currently under construction, offered one possible source for tritium, since it had not yet been operated and thus could be more easily converted. However, the GAC, thinking in terms of preliminary fusion designs needing large amounts of tritium, warned that the requirement could not be fully met simply by extending current methods. New slugs and chemical processing facilities, among other things, had to be rapidly developed and constructed if large quantities of tritium were required. Furthermore, better use of HEU-fuels could be achieved in a reactor with an entirely different conceptual design, moderated by heavy water.[15]

At issue for the Atomic Energy Commission was the short-term tritium requirement for the first tentatively scheduled test as well as the long-range production needs visualized for future thermonuclear weapons. In February 1950, the JCAE discussed four possible alternatives for producing tritium. One involved loading H reactor with enriched uranium for tritium production and using the still unloaded DR to make up for the loss in plutonium from H's diversion. The other possible approaches to tritium production were building six materials testing reactors (which Argonne National Laboratory was investigating), using a large linear accelerator currently under study at Berkeley Radiation Laboratory, or continuing design work on a heavy-water-moderated reactor similar to the one built at Chalk River, Canada, during World War II. The last alternative seemed to offer an efficient and realizable approach.[16]

The Joint Committee also focused on which reactors at Hanford to dedicate for the short-run, immediate tritium production. However, committee members quickly realized that tritium production at Hanford interfered with the production of fissionable plutonium. In response, Hanford investigated the possibility of using only slightly enriched uranium in one of the reactors, allowing for production of both tritium and plutonium. It soon became apparent that partial enrichment would necessitate the use of several piles in order to produce the needed quantities of tritium. Plutonium production would be greater if a single reactor were devoted to tritium than if tritium production were spread among many reactors. The H reactor became the choice for dedication to tritium because it could start producing tritium sooner than DR, whose waterworks was still under construction. Again, management decisions reflected the intersection of weapons policy, scientific constraints, and local practical matters.[17]

Compounding the anxiety about the thermonuclear program was the revelation in January 1950 that Klaus Fuchs had supplied detailed reports on weapon design to the Soviets. Fuchs was a young German physicist who had defected to Great Britain in 1933; he later worked at Los Alamos on the bomb and had stayed on after the war. At the date of his confession, he was employed at Britain's Harwell Laboratory, where he was the leading candidate for the post of research director. Evidence and his full confession revealed he had regularly supplied the Soviets with secret information through both their British and American espionage networks and that he had continued to do so even after the end of World War II. According to Generals K. D. Nichols and Herbert Loper, Fuchs's information significantly advanced Soviet capabilities in developing the Super. Senator McMahon warned that the Russians could bomb Washington or New York "a hell of a lot sooner" than the United States had originally expected. Strauss agreed that a greater degree of urgency was needed in the thermonuclear program and that "more bolts and locks" were necessary to ensure security. Fuchs's confession substantiated the concerns of the security-conscious and, together with the Soviet development of the atomic weapon, clinched the arguments both for the Super and for increased plutonium and tritium production.[18]

With these events suddenly converting the American monopoly on the winning weapon into a neck-and-neck nuclear arms race, the AEC moved directly to discussions on exactly how to run Hanford's H reactor to produce tritium, a new process that had not been fully tested. American scientists had not extensively studied highly enriched uranium fuel rods, so Commissioner H. D. Smyth suggested the use of the cooperative relationship with the Canadians to test the rods in the NRX heavy-water reactor at Chalk River. Heavy-water reactors were usually fueled with HEU, and thus, the Canadians had pertinent experience. As Smyth pointed out, the United States would obtain the necessary technical data much faster by cooperation than by separate research.[19]

By the spring of 1950, the AEC had decided how to approach the future production of tritium, both in the short term and the long term. In order to meet immediate testing requirements, General Electric started producing quantities of tritium at Hanford by dumping the regular slugs from H reactor and replacing them with HEU slugs and lithium deuteride target slugs. Based on the information gained from these trials, GE then planned to start an interim program for making stockpile amounts of tritium by loading both H and DR reactors with enriched material. However, for the long term, the Commission decided to produce tritium using heavy-water reactors located at a site other than Hanford.[20]

New Reactors at a Second Site

The Commission decided to construct two full-scale production reactors, moderated with heavy water, at a completely separate site, yet to be decided. Even though facing such uncertainties, it could make a few preliminary decisions rapidly, particularly the choice of contractor. By 1950, Crawford Greenewalt had moved to the rank of president of Du Pont, bringing the wartime experience with the Hanford reactors to the highest level of decision making in the corporation. The AEC decided, without much debate, to work with Du Pont as the contractor for the new site.[21] The Commission went ahead with the obvious choice, engaging Du Pont to do preliminary planning out of funds already available.

Du Pont accepted the contract but insisted on a personal letter from the president urging it to do so, as proof that it had not sought the work.[22] Through August 1950, Greenewalt worked to get a special request from President Truman in the form of a letter, to be able to demonstrate that the company's reentry into the nuclear business came in response to a genuine national defense priority, established and confirmed at the highest level. Eventually, Truman complied with a brief letter. Then and later, company officials regarded this specific, personally signed request from Truman as of great significance, using it to explain company participation in the project and at least to imply a commitment to the project as a national priority, even though the letter came 6 months after the company became involved. The letter was not only cited, but it was photographically reproduced both in public relations documents and in submissions to congressional committees so that Truman's personal signature could be noted. Du Pont was following orders and could prove it, to the government, to stockholders, to employees, and if need be, to the general public.[23]

The selection of the site for the new reactors moved along quite smoothly, considering the potential such a process naturally possessed for generating delays. Over a period of several weeks in the late summer of 1950, Du Pont engineers narrowed the choices to a short list of 17 sites. By the end of November, a site on the Savannah River near Aiken, South Carolina, had been chosen.[24]

When Commission chairman Gordon Dean reported directly to the JCAE about the site selection process, he made it clear that the experts at Du Pont were entrusted with the decision-making power over the site. Yet he included with his testimony a press release prepared by the AEC that stressed a pattern in which the government's site review committee considered *recommendations* from Du Pont. The difference in tone between the public statement and the statement to the JCAE, while one of emphasis, may have reflected a concern that the press and sectors of the public would not approve of too central a decision-making role for Du Pont; by contrast, the congressional committee, knowing the corporate record from the war years, might tolerate Du Pont's leadership on the issue.[25]

Under General Groves, military staff members and contracted experts made the decisions. In the world of the 1950s, the public tried to participate somewhat more directly. In contrast to the stormy controversies that erupted a decade later over corporate decisions on commercial reactor siting, the Savannah River site selection process was painless and only slightly more exposed to public scrutiny than the earlier siting under MED.[26]

A few public rumblings about the site choice process began to reach Congress, but they were readily handled. The complaints did not reflect objections to the risks that might be associated with the site, as in the later power-reactor disputes, but arose from the discontent of citizens and representatives of a variety of other sites who, with an eye to economic benefits, wished that the Commission had selected their own locations. Dozens of communities, discounting any unusual risks associated with a production reactor site, clamored for consideration. The JCAE responded to the advocates of different localities by reviewing the site selection process after it had been substantially completed. The Joint Committee agreed that the Atomic Energy Commission's choice had in fact been in the government's best interest, stressing that careful consideration had been taken of a number of objective factors, such as water temperature, military security, land cost, and minimum numbers of people to be displaced.[27]

The new reactors could provide a steady supply of tritium as well as more plutonium production capacity for the expanded stockpile and a backup facility in case of attack or accident at Hanford. The ideal new reactors would use highly enriched uranium as fuel and heavy water, or deuterium, as a moderator. Although a heavy-water-moderated production reactor had been recommended as early as August 1943 by the P-9 committee at the Met Lab, Groves had not authorized detailed planning for such a reactor during the war.

As far as the AEC was concerned, the Russian bomb announcement "crystallized" Commission thinking about a new round of production reactor construction, both of new design and of the old graphite-moderated design. In June 1950, as the new reactor planning and site selection proceeded, the North Korean Army drove across the 38th parallel into the Republic of Korea. With Soviet espionage, the Soviet bomb, and the Korean War in the news, World War III seemed imminent.[28] Following this crisis, the Atomic Energy Commission decided in October to add three more heavy-water reactors to the two already planned for Savannah River. In 1951 and 1952, as the war continued, the Commission added another expansion round of three new graphite-moderated reactors to be built at Hanford for the purpose of simply increasing the standard atomic bomb stockpile.

Even as the arms race heated up, the Commission anticipated eventual overproduction of plutonium. Commission chairman Gordon Dean and others could see by 1952 that the new Savannah River reactors together with the new Hanford reactors would allow for a high level of plutonium production that would produce, in only a few years, a quantity in excess of any conceivable anticipated military need. Although tritium had a 12.3-year half-life, and thus would require a steady production to keep up with the decay of slightly more than 5% of any stockpile in a year, the 24,000-year half-life of Pu-239 meant that the stockpile of that strategic material would be consumed only in weapons tests or actual military use. While some degree of permanent tritium production capacity had to be maintained, there would be no continuing need for plutonium production after a certain point. It was difficult to predict in 1952 exactly when that point would be reached, but Dean, in a thoughtful draft position paper, estimated that the surplus would arrive by the middle or late 1960s. Events proved him right.[29]

Knowing that a surplus of plutonium would be achieved in a few years made for a number of important considerations. It would then be possible to close most of the reactors and to reserve one or two of them for future tritium production alone. With that in mind, it was not necessary to build all of the reactors with long-term life expectancies in excess of 20 years nor with a permanent commitment to plutonium production capacity. But even if reactors were not built for durability, the imponderables of designing them in such a way as to allow production of either plutonium or tritium, or both in various mixed proportions, needed considerable thought and planning.

Urgent Schedules

During World War II, the first three Hanford reactors had been sited, designed, engineered, built, and brought to operation between the fall of 1942 and the winter of 1944/45. From the first letter contract with Du Pont to the operation of B reactor, the first reactor, was a period of 21 months; D and F reactors were operating within 27 months. The construction of the first postwar Hanford reactors moved even more rapidly, partly because of the speed of replicating existing design, partly because of experience gained in questions of control, slug handling, and river-water cooling. The new generation of reactors at Savannah River, with their entirely new designs, took longer to build, on average, but only slightly longer (Table 2). The Savannah River project was conceived in 1950; the first reactor, R, was begun in June 1951 and completed 25 months later, in July 1953.

Operation of the Hanford reactors generally began within a month or two after completion; the Savannah River reactors usually required a longer period of testing and minor modification before operation. By Hanford standards, the 38 months from start of construction to operation for C reactor at Savannah River was quite slow. However, by the standards of a later generation of nuclear engineers, such a pace would appear incredibly rapid, even reckless. The placing of R reactor in operation in December 1953, when the conceptual design had only been sketched out in December 1950, seemed to later nuclear specialists a remarkable achievement in engineering and management. Engineers of the 1980s and 1990s attributed the relative rapidity of the earlier generation's work to the absence of environmental legislation, public involvement, and an adversarial political atmosphere. There is no doubt that those factors complicated the life of engineers in the later period, but the successful and rapid work of the 1950s derived from a number of other factors that bear close examination.

The prompt schedule was possible because of several distinct, but related reasons, not immediately obvious in retrospect. The Korean War, the possible imminence of World War III, and a national consensus that sacrifices were required to stop the Soviets, all created, at the national level of the Atomic Energy Commission, an atmosphere of urgency and commitment. In that environment, Du Pont, the contractor with the most pertinent recent wartime experience, was the ideal choice.

As a contractor, Du Pont not only brought considerable reactor-building experience, but also particular methods and style, its particular corporate culture, to the task. Under MED, the corporation had worked quickly to design a reactor using only some fundamental concepts from the Met Lab, sometimes making arbitrary choices between alternatives and options under the pressure of time and with little chance to consider all the consequences of a choice. Reflecting its experience in private-sector chemical engineering, the corporation was quite capable of sorting through difficult design decisions. Du Pont efficiently resolved design choices internally by a system of checks and balances between its own divisions and departments. Du Pont handled its liaisons with other parts of the growing nuclear establishment with a minimum of bureaucratic delay, arguing successfully with

Table 2. Postwar production reactor completion schedules

Reactor	Date approved	Date started	Date completed	Months of construction	Date of operation
Hanford					
H	Nov. 1947	Apr. 1948	Oct. 1949	18	Oct. 1949
DR	Nov. 1947	Dec. 1947	Oct. 1950	34	Oct. 1950
Average start to completion, Hanford				26	
Savannah River					
R	May 1950	Jun. 1951	Jul. 1953	25	Dec. 1953
P	May 1950	Jul. 1951	Oct. 1953	27	Feb. 1954
L	Oct. 1950	Oct. 1951	Feb. 1954	28	Jul. 1954
K	Oct. 1950	Oct. 1951	Jul. 1954	33	Oct. 1954
C	Oct. 1950	Feb. 1952	Feb. 1955	37	Mar. 1955
Average start to completion, Savannah River				30	

Source: AEC 1140, "History of Expansion of AEC Production Facilities," 16 August 1963, 48, DOE Archives, RG 326, Secretariat, box 1435, folder I&P 14, History.

AEC procurement officials that emergency conditions should allow for noncompetitive purchasing of key components. Good scheduling, spurred by a sense of urgency, allowed planning and design work to be done even during construction of already-settled components, probably the greatest single contributor to rapidity and efficiency. From the selection of the site to the settling of literally hundreds of major and minor design and construction questions, Du Pont used its own flexible methods to good effect.

Participation in the Design Process

On technical design matters, the AEC relied very heavily upon Du Pont, although it also got input to varying degrees from other institutions in its now far-flung complex. The fundamentals of the conceptual design had been worked out by Walter Zinn at Argonne. AEC records show consultations not only with Zinn's reactor group, but with other groups at Oak Ridge, the GE-operated Knolls Laboratory at Schenectady, and the Canadian facility at Chalk River. By 1952, Argonne operated six programs in support of the Savannah River project, for a total of $2.7 million planned for the 1953 fiscal year. Knolls ran two programs at $2 million, and Oak Ridge conducted $400,000 worth of studies on separation of products, for a total of slightly over $5 million at the three sites in the single budget year. All three facilities engaged in training personnel for Savannah River, the high point being 229 trainees at one time in the first quarter of 1953. In contrast to the hundreds of millions spent through Du Pont corporation for the construction of the Savannah River reactors, however, the work at the other facilities was a minor part of the total.[30]

Du Pont, of course, paid for concrete and steel, not just studies, making its expenses much higher than those of the design consultants. Nevertheless, literally hundreds of the detailed technical decisions that gave physical shape to the technology were made by Du Pont engineers without the sort of direct military oversight and policy monitoring that Leslie Groves had exercised. Groves had worked through the Met Lab and blue-ribbon groups of physicists during the war years. In one 1950 throwback to the earlier use of the wisdom of renowned physicists, the AEC consulted with Eugene Wigner and John Wheeler, veterans of the MED effort. The old disputes between Du Pont and the Chicago physicists appeared to be forgotten as Wigner applied his enthusiasm for the heavy-water design to some of the early planning. Such participation was isolated and represented the exception rather than the rule in 1950, as the Commission relied on Du Pont to provide design coordination.[31]

A stark contrast between the MED style of managing the design phase and that of the AEC was the almost complete absence of outside consultant checks or restraints on the technical decisions made by Du Pont in the later period. During World War II, Greenewalt had been in constant communication with Groves and the Met Lab scientists over both major and minor details of design and operational problems, as in the resolution of the xenon-poisoning issue when B reactor first started. In the period 1950–53, under the civilian management of the AEC, Du Pont worked rather differently. The overall conceptual design; the scale of the reactors; their eventual product-mix between plutonium, tritium, and polonium; and of course, the provision of funds to build the reactors—all were decisions made by the Commission. But on vast numbers of smaller, yet important, practical design decisions, Du Pont appeared free of the type of external oversight exercised during the Manhattan project by Groves and the Met Lab scientists. These conclusions and decisions were reached through a Du Pont cultural style that its officers and engineers referred to as "flexibility," a term which accurately described its procedures.

Design Flexibility

As Du Pont began reactor construction planning in October 1950 even as the final narrowing-down of the site choices proceeded, engineer A. E. Church of Du Pont's Atomic Energy Division stated that the design team placed a "large premium" on "flexibility in the ultimate design."[32] This flexibility was a key to understanding the whole approach of Du Pont to the Savannah River task. Du Pont kept a large number of design choices open and allowed for several distinct types of flexibility on the project. At the simplest level, flexibility meant allowing the engineers to get moving with some design choices even while awaiting final design decisions. Du Pont staff resolved the tension between the requirement for early on-line production and the need for time for the best design by a well-thought-through process of temporary postponement of some decisions. Even as the plans were set on individual aspects, other, fundamental design questions were left open for discussion of competing alternatives. In October 1950, for example, Church deferred a decision on whether the heavy-water coolant and moderator should flow upward or downward through the core, but he expected prompt and prior settlement on such issues as lattice arrangement, moderator purification, monitoring, control rod positioning, and a gas envelope system.[33]

In the October discussions, Church set out an 11-point scope of work for the design division of Du Pont's engineering department. The scope of work called for preliminary pile design data: a general description and then details on tanks, fuel, lattice, control rods, monitoring, moderator purification, gas system, shields, charging and discharging, and materials to be used.

The AEC had decided that all the production reactors to be built at Savannah River would be scaled at about 300 MW and that they would follow the same conceptual design, heavy-water moderated and heavy-water cooled. The Commission selected an Argonne design as the one of several proposals best suited for further development.[34] The "basic concept," Du Pont engineer R. M. Evans later remembered, "had been developed by Wally Zinn." The basic experimental information on reactor physics and engineering had been developed at Argonne over the course of 1950, so that by December of that year a scope of work was spelled out that allowed a reactor to produce both tritium and plutonium or to produce plutonium only. Another objective was to be able to increase the power level through enriched fuel loadings. In effect, the scope of work defined what Evans called a "multipurpose reactor," one that could operate efficiently with various mixes of product and fuel. Thus, one meaning of flexibility was the ability to build a reactor without presetting its final specific product mix or its final power level.[35]

Even though the conceptual design came from Argonne, thousands of engineering details, from the concept to the final device, were left to Du Pont. Du Pont staff produced studies and reports on such features as control actuator design, the use of zirconium-clad thorium control rods, the removal of scale in heat exchangers, water cooling, shielding, and safeguards.[36] Du Pont designers did make use of pertinent research from Oak Ridge, Argonne, Knolls, and Chalk River. Nevertheless, the company was by no means simply carrying out designs developed by the scientific laboratories; rather, isolated pieces of the scientific, experimental, design, and engineering tasks were farmed out, with the bulk of the detailed work of all types being done by Du Pont personnel. It seemed the Commissioners established control only by setting the conceptual design and the general parameters; within those parameters Du Pont made almost all the detailed choices with the help of a scattering of AEC-paid contractors at other facilities.

With Zinn's heavy-water concept in hand, different teams of Du Pont engineers worked on four separate layouts, or configurations, of equipment simultaneously, in order to expedite the decision as to the most efficient arrangement. Church reviewed all four preliminary arrangement drawings provided by the design division, commenting on questions of space requirements, charging and discharging arrangements, the need for protection against bombing and earthquakes, and other details of the so-called "105 building," which represented the generic design for all the planned Savannah River reactors.[37]

In addition to the fixed decision to use heavy water as the moderator and coolant, the first scope of work also set down the "arbitrary" decision to use slugs as the form of fuel, "since no other form had been developed." Even though racks of fuel plates or other shapes might be more efficiently cooled in the heavy water because of higher volume-to-surface ratios, the background of work with canned slugs at Hanford provided the designers with a known starting point. In this decision, convenience and preexisting convention prevailed.[38]

Changing from fuel in slugs to other possible fuel configurations held out the hope of upgrading the power levels in the future. Since the factor limiting the power level was the internal temperature of the fuel element, designs that permitted more efficient cooling allowed for much higher

Du Pont used flexible engineering methods to design and to build Savannah River's R reactor. (U.S. Department of Energy)

power. But all the later designs were constrained by the tubes that held the original slug design; waffles, plates, and other noncylindrical overall configurations were eliminated very early. The design constraint imposed by this early decision to use slugs did not prove disastrous, however. Before the startup of R reactor, Evans noted, "We have flown considerably higher than the 700 M.W. figure in some of our optimistic guessings for which little basis of fact exists."[39] Later upgrades, in fact, took some of the reactors over 2,000 MW, so the early optimism was, in retrospect, quite conservative.

As Du Pont moved towards refining designs and beginning construction in 1951, the company's approach reflected its experience both at Hanford and in chemical engineering more generally. The intricacy of the work and the rapid pace left a tangled trail of memoranda, plans, committee reviews, and individual commentaries on choices. In this welter of communication, some patterns emerged. Heavy construction and some auxiliary building and infrastructure work could go ahead immediately, with postponement of detailed mechanical components for various periods. Du Pont executives used their system of postponing some decisions and reaching others promptly to allow for experimentation, revision of plans, and the pursuit of efficiency and the most rapid attainment of construction schedules. Early choices constrained later choices, and decisions were taken selectively with an awareness of how choices narrowed future alternatives.[40]

Early in the project, Du Pont installed a water treatment laboratory and semiworks to look into the effect of Savannah River water upon heat exchanger performance. Researchers were surprised to discover that intermittently chlorinated raw river water, "mud and all," was a better coolant in the heat exchangers than the same water treated by the expensive processes of flocculation and filtration. These semiworks experiments led to eliminating four costly water treatment facilities, one for each remaining reactor, before they were built. Further, the deionized pure heavy water as primary coolant proved far less corrosive of aluminum than Columbia River water had been at Hanford, even at much higher temperatures. Such experiments allowed redesign, sometimes with a cost savings, as the later facilities were being built.[41]

When confronted with preliminary AEC guidelines on radiological safety, J. E. Cole of Du Pont's Technical Division of its Atomic Energy Division suggested limits to the proposed policy. It was Du Pont's intention, Cole pointed out, to design so that "all normal effluents . . . will be well within the tolerances numerically defined." However, he pointed out, "we cannot guarantee that under unusual or unforeseeable circumstances, these tolerances may

Savannah River's reactors used heavy water both to moderate and cool the nuclear chain reaction, requiring extensive pipes to operate and control the system. (U.S. Department of Energy)

FLEXIBLE DESIGN AT SAVANNAH RIVER | 83

not be exceeded." He suggested a number of changes in the wording of the guidelines that made it possible to meet them. In effect, he suggested changing the regulations to conform to what he thought was possible, rather than trying to change the practices to what he believed were unworkable guidelines. In particular, he objected to the concept that discharges to ground or to water should not lead to contamination of possible future drinking water. In light of the fact that the company would be dealing with "radioactives whose half-lives approach 20,000 years, it is impossible on the face of it to produce 'demonstrable evidence' that some water contamination will not later occur."[42]

Cole pointed out that "one has to be practical about this sort of problem." He noted that "no substantial human action that modifies the earth's crust can be demonstrated in advance not to cause difficulty to later generations." In line with these thoughts, he suggested modifications that kept open such options as ocean dumping of radioactive waste and that modified the possible long-term legal ramifications of the early proposed guidelines.[43] To an extent, the Atomic Energy Commission was beginning to recognize a civilian-based set of priorities and a responsibility to later generations; Du Pont executives, ever practical, did not want such concerns to hamper design decisions and to prove unnecessarily restrictive. Cole informed the AEC that its guidelines were improperly worded and could not be considered logical in their existing form. While a later generation found it easy to condemn such an approach, Cole's corporate self-assurance on this score did not appear unusually arrogant or atypical in 1951.

Expeditious Procedures

The company's style of assurance and independence of operation was reflected in many ways. As the company planned a detailed program of experimentation in 13 areas, ranging from control to instrumentation, shielding, and reactor tank construction, rapid liaison with subcontractors and suppliers became essential. Du Pont explicitly indicated that to proceed expeditiously would "necessitate departure from established procedures, such as the elimination of bidding on equipment." Du Pont cooperated with General Electric at Hanford, as well as farming out parts of the project to Du Pont subdivisions and relying on programs at Argonne, Knolls, and Oak Ridge.[44] In effect, Du Pont officials let the government know that it expected special treatment regarding procurement because of the unique nature of the project. There was nothing sinister or particularly collusive in this approach; rather, it was a straightforward concern with moving ahead in a practical and nonbureaucratic fashion. General Electric had developed slug design and slug-handling tools, for example, working on the Hanford reactors, and it would have been foolish to put out for bid requests to supply those pieces of equipment, since no other company was working in the field.

As early as November 1950, Du Pont worked directly with General Electric—Hanford, asking that Hanford people test some newly designed aluminum cans and rods. Only after the work had been discussed by telephone and in writing with the manager of the Manufacturing Division at Hanford did the Du Pont managers work with the AEC regional officer at Savannah River, Curtis Nelson, involving him in "making the necessary arrangements with the Hanford Works to have this test scheduled."[45] The AEC was brought in to provide the formal financial and bureaucratic paperwork after the technical details had been settled. Some of the midlevel people working for General Electric at Hanford were former Du Pont employees; parts of a personal network remained in place despite the shift of managing contractor.

While such arrangements did not conform to any strict procurement protocol that required competitive bidding, it was the clear and practical way to get the government's business done. By the use of such day-to-day old-boy networks, Du Pont was able to work smoothly and quickly to achieve design and engineering progress at a pace considered phenomenal 30 or 40 years later.

Du Pont personnel understood the necessity for control and coordination of their work and were experienced in sorting through the pride of ownership that generated advocates of one system over another. With such issues in mind, Du Pont personnel established internal checks and balances between their Atomic Energy Division (AED) and their Engineering Department. Within the AED, a further degree of internal checking and control existed between the Technical Division's own Reactor Physics and Reactor Engineering sections. Because these internal review levels allowed for a check between competing groups, Hood Worthington, of Du Pont's AED Technical Division, took exception to suggestions from outside consultants that there should be some sort of outside review of the control rod design system. In his view, Du Pont had done a great deal to ensure that design decisions were reviewed and re-reviewed; the establishment of another, external review seemed quite redundant to him, and he urged the AEC to accept Du Pont's arrangements. In general, company engineers believed Du Pont had the experience and depth of personnel to sort through all of the alternatives with an objective resolution of competing views. Information, reports, ideas, and experiments provided by others at Argonne, Knolls, Oak Ridge, and elsewhere were taken as advisory, not controlling, and folded into the internal Du Pont decision process.[46]

Postponed Decisions

Early considerations affecting the issue of flexibility emerged in 1951, as Du Pont engineers began to set some designs more firmly while postponing other decisions. In particular, the Atomic Energy Commission continued to leave open the issue of what proportion of the reactor work to devote to plutonium production and what proportion to tritium production. Consequently, Du Pont had to design in an optimum fashion that allowed alternate fueling schemes. In addition, early in 1951 Du Pont kept plans open for the production of polonium, and designers had no clear policy guidance as to the specific proportions expected of each of the three products.[47]

In 1951, Du Pont reported to the AEC that it was keeping open the relationship between plutonium and tritium production and noted a recent discussion by the Commissioners suggesting emphasis on plutonium production. The first two piles, based on such hints but not on firm policy, were set up for plutonium as the higher priority. Du Pont designers chose the practical way to implement the Commissioners' policies as they were decided and then let the regional managers of the AEC know the choices they had made.[48]

Keeping the options open for as long as possible proved good business. The Commission delayed a decision on the issue of product mix until R went into operation in November 1953 and P and L were under construction. The Commission then ordered Du Pont to alter the design of L to allow for charging with highly enriched uranium and producing the maximum of tritium without regard for plutonium production.[49]

Similarly, the AEC did not reach an early decision on whether the heavy water needed to be refrigerated or the extent of future power upgrades, and Du Pont willingly designed around those issues as well. In the first year of construction, Du Pont reported that the company held open flexibility on a wide range of issues: emphasis on plutonium over tritium, the possibility that the reactors might later be upgraded to twice the original rating, and future alternate fueling schemes.[50]

Rather than expressing frustration or demanding decisions, Du Pont designers accepted as good engineering practice the system of designing those elements that had been decided and postponing those elements that needed further study, the same corporate style which they had used at Hanford, and which can be traced to their earlier developmental work with nylon.[51] With a practical but flexible layout of major components, some could be worked on while the options on others were discussed and narrowed.

The interplay of factors on such flexibility became more intricate through 1952 and 1953 as more and more decisions had to be made and actual concrete and steel had to be set into place. For example, in designing the control rod actuator system, leaving open the final decision as to required positions for the control rods while designing the system to move the rods presented difficulties. One designer noted, "The major reason for complexity and cost of the present design is that we have asked for an unusual degree of flexibility in a mechanism which involves so many elements." The control rod servomechanisms had to be designed for possible future changes in the ganging of control rods into clusters. As the control rod design moved ahead, no single document contained all of the specifications, further complicating the work of the subcontractor, American Machine and Foundry, which had been chosen to put together the control rod systems. The engineer who described all of these problems remained a staunch advocate of the design as it had evolved, in the face of "adverse criticism."[52] Flexibility led to complexity as a fact of life, and the designers worked with it.

Du Pont designers worked with two fundamental classes of flexibility. One class included choices postponed until the best design could be determined either by experimentation or by further discussion. The second class included options built in, in order to deal with future policy choices. Although similar, one type of flexibility represented a postponement or delay of *design* decision, while the other was a design decision implemented to accommodate future *policy* decisions currently held open. Despite so many demands for different kinds of flexibility, Church had been able as early as 23 October 1950 to provide 19 pages of single-spaced text detailing choices that had already been determined.[53]

One early firm decision, arising from the conceptual design, was the unique structure of the cover of the reactor vessel, a laminated steel plate 19 feet in diameter and 4 feet thick and weighing about 100 tons. This cover plate, or "plenum," was drilled with more than five-hundred 4-inch tube holes, set on 7-inch centers. This elegant piece of metal, the designers recognized from the beginning, presented "an unusual task of handling, fabricating, machining, and shipping." The contract for the work was placed with New York Shipbuilding, of Camden, New Jersey, which produced not only these pieces, but the vessels and much of the primary piping as well. As the work proceeded at Camden, a scrupulously complete photographic and narrative record of all the fabrication was maintained, with a view to leaving guidance for those who might attempt "an identical job in the future." When a choice was made to take on such an intricate and difficult job, it was not only a firm choice, but it was also documented step-by-step so that later, if desired, the fabrication could be replicated, down to the last fraction of an inch.[54]

The concept of going ahead with planning and with setting firmly in concrete and steel certain features while delaying fundamental decisions on other features did not characterize later production reactor planning through the 1980s and 1990s, when preliminary design requirements documents reflected a method of settling on many more design choices on paper. A later generation of nuclear engineers might find the 1950s Du Pont principle of retained flexibility during actual construction anomalous; that degree of flexibility was made possible by the relatively free hand provided to the corporation by the government. Further, it reflected the urgency imposed by attempting to achieve production as soon as possible while working out design issues. As a chemical engineering firm, Du Pont had extensive experience in constructing plants and postponing the resolution of various components while proceeding with others. Nuclear reactor decision-making in later decades reflected the preplanned rather than the flexible approach, a very basic change in engineering style.

For Du Pont, the machine was part of a system used to make a product. As a chemical firm, Du Pont had no difficulty considering the reactor itself a flexibly designed machine with several possible functions; later, it redesigned the plant, as necessary, to produce the products demanded by customers.

Despite the impact of Du Pont's chemical engineering background and specific corporate style on the design of the Savannah River reactors, the emerging profession of nuclear engineering was at that very time drawing heavily upon several separate streams of engineering tradition and, as a profession, developing a leadership dominated by experts from electrical engineering. At the American Institute of Electrical Engineering (AIEE) meetings in 1953 and 1954, the president of the organization, Donald A. Quarles, sought to give more definition to the emerging field of "nucleonics." Quarles, an electrical engineer himself, had been chief of the Western Electric subsidiary operating the Sandia Laboratory, which was involved in nuclear weapon manufacture, for the Atomic Energy Commission. Quarles worked closely with Walker Cisler, president of Detroit Edison, and with

In 1951 Du Pont contracted with the New York Shipbuilding Company to construct the vessels for the early Savannah River reactors, which were scrupulously measured for accuracy throughout construction. (Hagley Collection, U.S. Department of Energy)

T. Keith Glennan, another electrical engineer, in establishing the Atomic Industrial Forum (AIF).

Quarles, Glennan, and Cisler, who would play major roles in lobbying for nuclear power research over the next decade, emerged very early, in the 1950s, in leadership and policy-creation positions in the field. Quarles would go on to positions in the Department of Defense at the Secretarial level; Glennan, who served on the Atomic Energy Commission, remained a major senior statesman of the field; Cisler headed the Atomic Industrial Forum. Even so, the contributions of electrical engineers were primarily in only a few areas of reactor design, particularly control apparatus. The emerging discipline, despite its domination at the policy level by electrical engineers, represented a fusion of several different streams of engineering know-how, especially from the fields of chemical engineering and mechanical engineering.[55]

Long-Range Concerns

Planning for the first of the new heavy-water reactors began in 1950; by 1955 all five were completed and operating. Although constructed along the same lines, K and C reactors included innovative ideas worked out on the first three (R, P, and L). The flexibility in the design and construction process allowed for later adaptation in target elements, in production, and in safety that enabled the longest-lived of the reactors, K and C, to operate into the 1980s and allowed the rebuilding of K to meet standards in the 1990s.

However, a number of fundamental issues began to surface in the first operation of the Savannah River reactors. First of all, the reactor operators employed by Du Pont found routine operation a tedious duty, and in order to engage the company's best personnel, Du Pont had to construe the work as involving a continued program of innovation. While innovation in target elements, in safety, in isotope production, and in application of computer methods proceeded, the groups involved in the work had difficulty adapting to the rigorous demands of routine production.

Second, as the reactors aged and went through rebuilding and redesign, managers grew increasingly concerned about safety, both of workers and of the general public. This concern increased partly because of growing public awareness of the hazards of nuclear reactors and partly because of the emergence of internal experts who disagreed over the interpretation of the seriousness of the variety of incidents. Some Du Pont executives and engineers began to be concerned that any accident, even if technically minor, put the reputation of the corporation at risk in a public relations and political sense, if not in a legal sense.

Third, as the reactors aged and as stress-corrosion cracking along welds produced minor leaks, questions as to the eventual life span of the reactors began to demand attention. There was no "design-basis" life expectation, and both AEC and Du Pont officials avoided any direct reference to such expectations. However, safety officials at Savannah River referred to the fact that the reactors were getting older as early as 1961.

Fourth, the unique heavy-water design of the reactors at Savannah River eventually led to another problem, not apparent at first. In 1955, when all five reactors were operating, there was no commercial reactor industry in the United States. By 1966, 27 power reactors had been ordered. By the 1980s, about a hundred had been built.[56] Almost all the power reactors used either pressurized or boiling water for cooling and moderating. Only one experimental model, the 17-MW(e) Carolinas-Virginia Tube Reactor at Parr,

South Carolina, was heavy-water-moderated and -cooled like the Savannah River reactors. This meant that the state of the art of reactor operation at Savannah River grew and changed in considerable isolation from the practices and methods in the burgeoning reactor industry. As time went on, that technological and cultural isolation became more pronounced. Specifically, Savannah River technicians were slow to emulate new methods of risk assessment developed in the commercial sector. Hanford was similarly isolated because of exemptions from the safety rules applied in power reactors. On the whole, engineers at both facilities rarely attended meetings organized by the emerging profession of nuclear engineers; increasingly, production reactor engineers became intellectually and institutionally insulated from the newly defined mainstream of nuclear engineering.

Despite its problems and its isolation from the culture of power-plant-oriented nuclear engineering, Du Pont had moved quickly and responsively as the United States entered a nuclear arms race with the Soviet Union. Within a 5-year period Du Pont had designed, built, and brought into production a whole new production reactor complex. The Savannah River reactors were innovative and effective. Despite early and continuing concerns with safety, and despite very closely held reports of dangerous but minor leaks, the reactors never experienced a major accident. The Savannah River complex continued to provide tritium for the nation's nuclear stockpile through the vagaries of the Cold War, over a period of more than three decades.

5 • The Arms Race Arsenal

Over the decade of the 1950s the family of production reactors grew from five to thirteen, with another in planning by the end of the decade. The five built at Savannah River discussed in the previous chapter followed the conceptual design of heavy-water moderation rather than graphite moderation. The basic conceptual design of three new additions at Hanford emulated that of B, D, and F reactors, built during the war, and DR and H reactors, added in the early Cold War years. The incremental improvements that General Electric had made on the early Du Pont designs would become incorporated in the reactors built at Hanford in the 1950s.

The interplay of international relations, domestic politics, and disputes and tensions within the weapons complex between the different managerial hierarchies all contributed to the particular shapes of the new members of the production reactor family at Hanford. The choices defining the new reactors represented much more than modernization based upon experience. The three new reactors at Hanford were built under a revived wartime environment—that is, an intense international arms race. That consideration meant that the reactors had to be completed rapidly and that they had to be designed for higher power levels and higher production levels than the earlier models. Advances in understanding the effect of radiation on graphite and in knowing the effect of operating at higher flux allowed for new designs. Yet rapid construction could best be achieved by closely following earlier designs, a method incompatible with the goal of building on a new scale of power and incorporating the new scientific knowledge. Safety was also an issue. Demands for increased production created tensions between the Production Division of the Atomic Energy Commission, advisory committees of experts dedicated to safety, and General Electric as contractor.

The Production Division was in the difficult position of attempting to match the capacity of the weapons complex with tentative weapons "requirements" established by the Joint Chiefs of Staff and transmitted to the AEC through the Military Liaison Committee. In the early 1950s, the Joint Chiefs tended to structure each year's requirement as a percentage increase over the prior year, but by 1955, the Commission developed a more realistic planning method for requirements based on several factors. Balancing speed of construction, large-scale operation, high neutron flux, and safety all led to specific technological decisions in the effort to fulfill the requirements. All of these factors continued to leave a tangled trail of decision making, still largely hidden from public view.[1]

Korea and its Impact

In August 1949, the American nuclear monopoly had been broken with Little Joe. On 24 June 1950, following the withdrawal of U.S. postwar occupation troops from below the 38th parallel in Korea, the Soviet client state of North Korea launched a full-scale invasion of South Korea. Unlike the more gradual takeover of satellite states in Eastern Europe by domestic Communist groups under the protection of the Soviet Army, the North Korean attack was seen by President Harry Truman and the American people as a clear-cut case of military aggression by a Communist state against a democratic state. Many in America assumed that the end of the U.S. nuclear monopoly encouraged Soviet adventurism through its Asian satellite. Acting quickly, Truman committed U.S. air and ground forces to the defense of South Korea; a U.N. Security Council resolution gave the American response the legal character of an international police action.

The outbreak of the Korean War immediately deepened concerns about nuclear material production rates and the lack of a clear weapons lead over the Soviet Union. As the United States committed troops to Korea, the AEC and the JCAE discussed new goals for the Hanford site.[2]

In particular, Sen. Brien McMahon raised the point that new construction of Hanford-type reactors could supplement already existing plans for the two heavy-water reactors at the Savannah River site, which had been approved in May 1950. Adding urgency to the senator's concern were new intelligence reports that the United States might only have a one-pile advantage over the Soviets and might be losing any superiority in gaseous diffusion separation of U-235. With the Korean invasion, McMahon felt the time was at hand for a full reappraisal of production schedules. The emergency favored those who argued for the dedication of more resources to the weapons program.[3]

William Borden, executive director of the Joint Committee, only heightened McMahon's apprehension by arguing in a top secret report that failure to pursue both the heavy-water reactors and the Hanford-type ones exposed the United States to "unreasonable risk." It would make sense to build both, he thought. He argued that it was possible that the Soviets were well ahead of the United States in successfully developing a thermonuclear device, which in fact was the case. If so, then Americans needed some "insurance." Though the Atomic Energy Commission understood that the Hanford designs were obsolete and inefficient by comparison to the heavy-water design, the graphite piles did offer security as proven sources of fissile materials. It might be advisable to proceed with the tried and true, if dated, graphite design, rather than relying upon the untried heavy-water design for expanded production. As GE's director of research, C. G. Suits, pointed out, the only base-line for heavy-water engineering was the Canadian reactor at Chalk River, a small device that had encountered difficulties over the course of its lifetime. Since heavy-water reactors had to be designed as well as built and since the graphite-moderated H reactor had been built in the astonishing period of only 17½ months, it would make sense to get another graphite reactor under construction immediately to meet the goal of rapidly increasing production. In this way, questions of timing and the international crisis had a direct impact on the question of conceptual design choice. But final decisions on further Hanford reactors were not reached until the Korean War intensified.[4]

The New Round at Hanford

Further JCAE discussions in July 1950 brought out the multiple factors involved in deciding whether to build more reactors at Hanford. MLC chairman Robert LeBaron made clear that meeting tritium requirements should not be the only concern. The United States needed "sufficient flexibility" in its facilities to meet changing needs for components for *either* atom or hydrogen bombs. Devoting a Hanford reactor to tritium production for a year necessarily cut down the stockpile of plutonium, though AEC chairman Gordon Dean noted that the military had considered this situation and was not "exercised over the loss." Disruption of the "well integrated program" at Hanford was a factor to be considered against additional construction at that site. Adding one more reactor would require building another redox facility in order to continue retrieving U-238 otherwise treated as waste. Military advisors also continued to express concern that Hanford was becoming a vulnerable target. An accident or an attack could eliminate all production there, they feared. War planning at that time was predicated on possible Soviet over-the-pole aircraft attacks. Savannah River at this time seemed safer than Hanford, since it would be far more difficult for polar-route Soviet bombers to reach the site in South Carolina than the site in Washington State.[5]

In late October and early November 1950 U.S. forces started to encounter Chinese soldiers in North Korea. Soon afterwards, U.N. troops marched into a trap, as 100,000 Chinese "volunteers" came to the aid of their North Korean comrades. In response to the entrance of China into the war, President Truman called upon Americans to make a "mighty production effort," suggesting the degree to which Truman viewed the Korean War as a reprise of World Wars I and II. One answer to this call came from the Atomic Energy Commission, which ordered General Electric on 23 January 1951 to begin work on a sixth Hanford reactor to be built in the B area and called C. Commissioners noted that the new reactor was not "absolutely required" for meeting production goals, but it did offer added capacity.[6]

General Electric started designing the Hanford C reactor in March 1951, and construction got under way in June. Though still relying on the World War II reactor plans developed by Du Pont, GE introduced further modifications that provided greater overall production rates. One important step involved enlarging the plumbing facilities so that water flow could be increased beyond that of the original design. The more the fuel rods were cooled with water, the higher the power levels they could sustain, resulting in more plutonium or tritium. Another improvement related to the graphite-to-uranium ratio. When B, D, and F were built, the scientists and engineers did not know precisely what the physical constants of uranium isotopes were. As a result, they designed these first piles with a ratio of graphite to uranium as close to $k = 1$ as possible. By the time C reactor was built, designers knew there was some reactivity to spare, so they reduced the ratio of graphite to uranium in their reactor designs. This adaptation increased the probability of neutron absorption and promised higher production rates.[7]

Even as GE pushed ahead with building C reactor to meet future military requirements, Congress and the AEC began debating the need for still another reactor at Hanford. Weapons tests at the newly opened Nevada Test Site—including the Ranger, Buster-Jangle, and Tumbler-Snapper series—continued to demonstrate the feasibility of new, more efficient plutonium-weapon designs and consequently showed that the fastest way to guarantee a vastly expanded stockpile of weapons was to step up plutonium production. JCAE members began to question in 1951 if the production ratio of uranium to plutonium that had been established earlier would continue to prevail over the next decade. In addition to the Korean War, the growing perceived threat of a direct World War III between the United States and the Soviet Union contributed to the drive to increase nuclear production. Amid the uncertainties, building more production reactors would be central to nuclear security and to maintaining the ideology of deterrence.[8]

Jumbo Reactors

On 16 January 1952 President Truman decided on the increased ratio for plutonium over uranium-235 production and directed the AEC and the Department of Defense to develop programs in line with the new objectives. Both agencies had already worked on plans to increase plutonium production, and the next day they submitted to the JCAE a joint report that addressed the new requirements. On 25 February 1952 Truman approved their proposal to add new reactors to existing production sites and to build necessary support facilities. Originally, twin reactors at Hanford and a sixth heavy-water reactor at Savannah River were included in the plan, but by June 1952 the Commission determined that requirements could be met without the sixth Savannah River reactor. The consequence of Truman's 1952 decision, then, would be two new reactors at Hanford, in addition to the new C reactor already being built.[9]

As is reflected in Figure 1, the Hanford reactors were approved in two separate rounds after those at Savannah; the overlapping construction and completion schedules brought a total of eight new reactors into production over the period 1952–55.

K West (KW) and K East (KE), the newly slated reactors located at Coyote Rapids between the B-C and D-DR areas at Hanford, represented a transition for production reactors in several ways. They were larger and more powerful than the neighboring piles, ensuring production of weapons-grade fuel far into the future. At the same time, these Jumbo reactors, as they were called, demonstrated for the first time the concept of converting waste heat into productive energy for heating and cooling the buildings' work spaces.[10]

While the designs of the KE and KW reactors reflected the demand for increased production, they could also be construed as the beginning of an effort to harness nuclear energy for peaceful purposes. Still using essentially the same graphite reactor technology as had been employed in the original Hanford reactors, General Electric and the Commission designed the twin K reactors to handle power levels starting at 1,800 MW(thermal) (MW[t]). This was a huge increase from the original 250-MW[t] design level of B, D, and F reactors. Even the most recent facility, C, had been rated at only 750 MW when it started in the fall of 1952. All of these ratings appear deceptively high when compared to the ratings for early electric-power reactors built in the 1960s, which were posted in megawattage (electrical), a figure based on the much lower electrical *output* rather than heat-generation levels used to define the capacity of the production reactors.[11] Changes in the K water systems were crucial to the higher power level. Improved pump designs allowed Hanford to reduce the number of water pumps from 50, the number installed in the first piles, to only 18 while also increasing the amount of water being pumped fourfold. As a result, each reactor with its own water plant had an initial flow set for 125,000 gallons per minute and capability to increase to 140,000. Otherwise, the water facilities duplicated the layout of previous reactors, with water being pumped from the Columbia River through a filtration plant and a high-pressure pumping station to the pile. On exiting the pile, the water would cool in retention ponds, where short-half-life radioactive isotopes would decay, before the effluent was to be discharged to the river.[12]

Physically, the Jumbo reactors were more massive than previous reactors. They each used 2,800 tons of graphite, 1,000 tons more than before, to make a 41 × 41 × 33.5-foot irregular parallelipiped, roughly cube-shaped. A concrete shield was used instead of steel masonite. Slightly smaller lattice spacing between process tubes and a larger number of tubes represented further incremental modifications in the Jumbos.[13]

Functionally, KE and KW also departed from earlier models owing to their "dual-purpose" capabilities. Exit water from the graphite block was pumped to a heat exchanger, which transferred heat from the cooling water to an ethylene glycol water solution. The antifreeze solution then transmitted its newly gained heat to air ducts in the K reactor area, supplying heat

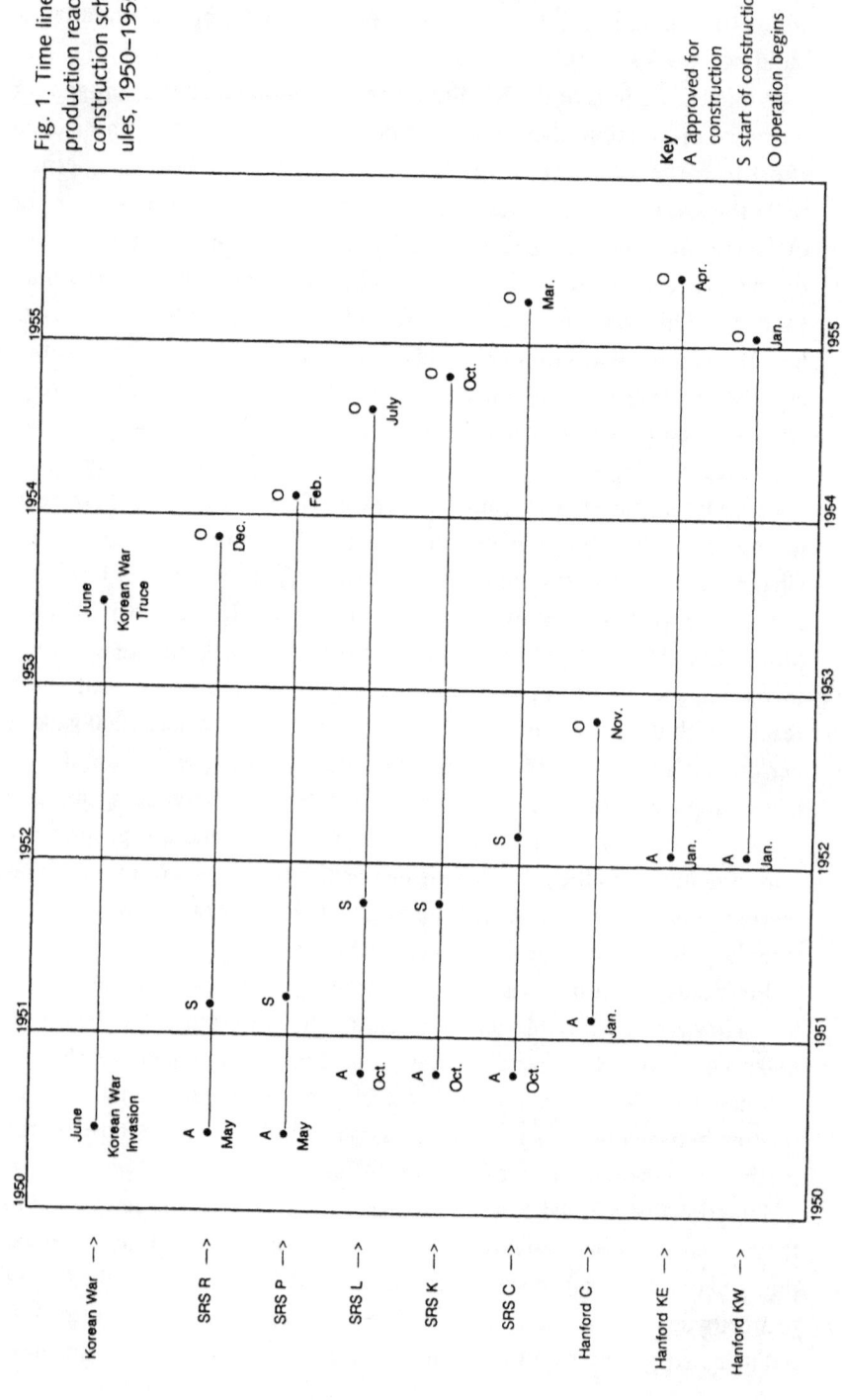

Fig. 1. Time line of production reactor construction schedules, 1950–1955

to the buildings. By keeping the pressure higher in the secondary ethylene glycol loop than in the primary loop containing radioactive water, General Electric and co-designers from C. T. Main, Inc., ensured that radiation did not travel from the cooling water to the heating system through any minute cracks or leaks but would stay in the once-through water coolant passed out through cooling ponds to the river. Through this simple process, KE and KW set a minor, but at least symbolically significant, precedent for various power reactors that would use heat exchangers to generate steam in the near future.[14]

Although General Electric took some pride in this innovation, AEC general manager K. E. Fields did not think the technology development was particularly dramatic. Indeed, GE had not departed from the original conceptual design of the World War II–vintage graphite-moderated, Du Pont–engineered piles in any significant way. Previous production reactors had the potential for this same application of heat generation, but economic considerations prevented its serious consideration. Fields attributed its use in 1955, when the K reactors first started operations, to the fact that reactor cooling water could then be heated to significantly higher temperatures than were permissible a decade earlier. More heat generation from the higher power levels meant the possibility of more economic use of that heat. His position was that heat could have been generated at any time and that it had become worthwhile trying to use it.[15]

In terms of cost, the Commission proclaimed in its annual report for the second half of 1953 that the K reactors' heating system would save an estimated 1.5 million gallons of fuel oil each year. In less than 8 years, the $614,000 investment for the specialized heat take-off equipment would be paid for through fuel savings, making the dual-purpose idea economically beneficial. These twin reactors also achieved cost savings from enhancements to the central control area. By operating the various process buildings through remote and essentially automatic control in a centralized area, General Electric saved labor costs. Each Jumbo reactor required approximately 300 people to run its operations, while H reactor, a far smaller pile, needed 400. Total operating costs for one K reactor, taking into account both the lower energy costs for heating the buildings and the reduced labor force, worked out to $1 million less than H reactor per year. With the truce in Korea in the summer of 1953, and with an active war no longer providing a justification for all-out weapons production, reduced cost at the K reactors was welcome news at the Commission.[16]

Balancing Safety with Production

As GE began operations in the newer production reactors, the Reactor Safeguard Committee (RSC) and its successor organization, the Advisory Committee on Reactor Safeguards (ACRS), established in 1953, both criticized the emphasis that the Commission's Production Division had placed on plutonium production over public safety in regard to the risk of major catastrophes. Both of the safety-monitoring groups defined safety as controlling or limiting the risk of *nuclear catastrophes*, placing far less emphasis on environmental hazards from *routine operation*. As contractor, GE found itself in a difficult position. At times, company managers sought to limit production in order to meet a safety objective, only to receive reprimands from the client, the Production Division. At other times, the RSC would refuse to endorse a practice adopted by the Production Division. Ultimately, when caught between the demands of safety and production, the company sought to find technical solutions that would allow safe production in the quantity demanded.

Due to the substantial difference in scale between the Jumbo reactors and other reactors at Hanford, GE and the RSC discussed ways to avoid potential dangers from a reactor designed to run at 1,800 MW. One involved development of a "comprehensive" startup program for the K reactors that preceded initial operations. In the case of safety systems, such as auxiliary process tube cooling and graphite wetting for handling a loss of cooling water accident, the two groups agreed that a slow approach was justified. On some other matters, the company and the RSC disagreed. In early 1953, the RSC suggested that safety devices be installed in the K reactors to warn of approaching criticality and to shut down a reactor in case of a loss of coolant incident. A. B. Greninger, the company's engineering manager for Hanford, sardonically reminded the RSC that neither instrument existed, nor was there any likelihood of their invention in the near future.[17]

One major incident drawing attention to the safety features of the K reactors occurred on 4 January 1955, when General Electric shut down KW during its startup operations due to a process tube water leak that appeared to be associated with a slug rupture. The reactor had been running for only 17 hours at low levels. After several days spent studying the affected area, company representatives reported that the tube and slugs had "melted considerably," indicating a major operating incident. GE and the AEC conducted an investigation that determined that cooling water had been blocked from entering a particular tube before operations began by a plug, which had been overlooked during startup. The problem was compounded by the fact that the pressure gauge for measuring flow through the process tube had been improperly set and calibrated. Supervisors had failed to notice either of these conditions when preparing KW for operations. It was noteworthy that the incident stemmed from two unlikely events compounding each other, rather than from a single major catastrophic design basis accident.[18]

GE and the RSC held different perceptions of risk and sparred over how to evaluate the reactor operations. The RSC focused on the perceived risk of a catastrophe and what its effects would be on the surrounding human populations and environment. In the view of the RSC, the magnitude of the risk increased correspondingly as conditions changed, such as raising power levels or using fully enriched loadings. General Electric disagreed. So long as the company modified its operating procedures to accommodate upgrades, General Electric did not believe either the "*probability* of an incident or the *magnitude* of an ensuing disaster" would increase.

The two views reflected two slightly different orientations, beginning to emerge in the 1950s, over the evaluation of systems risk. The RSC took a more traditional, deterministic approach, with a focus on a worse-case scenario, the means of avoiding that scenario, and the possible consequences of the scenario. It emphasized reviewing *safety devices* to effectively forestall potential catastrophes. GE took an approach that was beginning to be considered in the emerging profession of nuclear engineers, of defining risk as a combination of both the probability and the magnitude of an event. General Electric would attempt to modify *procedures* to hold the probability of an accident to a low figure even with the change to a larger scale of operation. The difference in practice might be slight, but it would lead to somewhat different emphases. Following the difficulties encountered in starting KW, the company had to admit that carefully planned operating procedures did not always eliminate the chance of an accident, when human error led to skipping a step. Such experience would suggest that it was not always possible to dismiss consideration of unlikely events.[19]

Solid fuel slugs could not withstand increased power levels because the hotter temperatures brought the slugs close to the boiling point of water; boiling would create steam voids and loss of coolant. The demand for increased production drove the technical search for new ways to reduce slug failure and to guarantee better cooling. A new slug design developed at Hanford by company employees in the late 1950s involved coring the center of the elements so that water could flow both through and around the rods, internally and externally cooling them. When aligned in the process tubes, the slugs created a continuous channel down the center through which cooling water could flow. With large-scale loadings in the reactors, GE personnel believed they could obtain "maximum power levels" with these internally and externally cooled (I&E) slugs.[20]

The AEC review panel, now named the Advisory Committee on Reactor Safeguards, cautiously approved GE's use of the I&E slugs. However, C. Rogers McCullough, the ACRS committee chairman, noted that from a safety viewpoint, there were "both advantages and drawbacks" to the new fuel elements. On the one hand, the I&E design promised fewer slug failures because of the greater cooling abilities, thus allowing for meeting the Production Division demand for higher production levels. Since slug ruptures imposed increased risks to reactor operations and also contaminated effluent water, decreasing the incidence of ruptures helped offset safety concerns over the higher power levels. However, McCullough also recognized that with fewer slug ruptures, the company would lengthen irradiation time, which could eventually bring the number of failures back up to present levels. In this case, McCullough prodded GE to continue developing safety improvements to match the power upgrades.[21]

McCullough's hesitation to grant outright approval to power upgrades sharpened following news of the October 1957 Windscale Pile No. 1 accident near Seascale, Cumberland, in Great Britain. The two Windscale graphite-moderated, air-cooled production reactors had started operations in 1950–51. On 7 October 1957, Pile No. 1 was shut down for a planned energy release, called a Wigner release, which was the cooling step following annealing. By 8 October, British operators believed that the graphite temperature was decreasing too quickly, so they restarted the pile.

In actuality, the temperature readings were not accurate—the reactor was much hotter than the instruments indicated—so the restart caused a major graphite fire. The added heat initiated a self-sustaining reaction that burned the graphite in an area encompassing 150 channels. Air cooling by convection or forced flow failed to reduce the temperature, and in fact only supplied more oxygen. Following other attempts to control the reactor, on 11 October, authorities flooded it with water, permanently destroying the pile. Substantial quantities of gaseous fission products had already escaped through the reactor stack. Although far less publicized and less catastrophic than the Chernobyl accident nearly 30 years later, the Windscale fire demonstrated to nuclear engineers some of the inherent dangers in graphite reactors and sent a shiver of concern through those who realized what had happened.[22]

Though recognizing that the Windscale reactors were substantially different in design from the Hanford reactors, Edward Bloch, the Commission's director of production, requested data on the expected temperature rise in graphite if all of the residual heat were released suddenly from the Hanford reactors. He also reviewed the adequacy of emergency plans in the event of a Windscale-type incident. O. H. Greager, GE's manager of research and engineering at Hanford, assured the Commission that though the reactors contained "substantial quantities" of stored energy, a sudden release was considered "impossible." Measurements of the stored energy in the different zones of the graphite block indicated that the rate of self-annealing had been sufficient to keep the stored energy at a safe level. Hanford reactor operations ensured that safe temperature levels were retained.[23] Nevertheless, the catastrophic accident at Windscale gave good reason to be concerned about the worst-case scenario.

Over the period December 1957 to February 1958, as GE employees prepared to load all of the reactors with the newly designed I&E slugs to achieve the higher power levels, the ACRS resisted. The safety committee may have been in "complete accord" that there were "no serious adverse nuclear effects" from these fuel elements. However, it could not ignore the cumulative risk from the power upgrades. In its opinion, the Hanford reactors were still "potentially dangerous facilities," especially in the event of a loss of coolant accident. The ACRS felt that the Commission, by running the piles at higher power, was accepting a greater degree of risk than in any other existing reactor.[24]

In this case, GE disagreed with the Advisory Committee, pointing to its cumulative experience at the Hanford site, improvements in instrumentation, increased knowledge of the production process, enhanced operator performance, improved maintenance, and rigorous procedures which, in the opinion of General Electric, decreased the likelihood of an incident even as the production levels steadily increased. In addition, if an accident did occur, the company argued that while the concentration of short-lived fission products would increase proportionately with the power upgrades, the hazards from long-lived fission products were determined from accumulated exposure and not from reactor power levels. Hence, in GE's view

Table 3. Power levels at Hanford reactors, 1958

Reactor	Year built	Design level (MW)	1958 level (MW)
B	1944	250	1,350
D	1945	250	1,350
F	1945	250	1,350
DR	1949	250	1,350
H	1950	250	1,350
C	1952	750	1,600
KW	1955	1,800	3,000
KE	1955	1,800	3,000

Source: AEC 1140, "History of Expansion of AEC Production Facilities," DOE Archives, RG 326, Secretariat, box 1345, folder I&P 14, History.

of risk, the consequences of the accident were not a function of the power level of the reactor. Power levels, in the company's opinion, did not determine the danger to personnel on the site.[25]

Despite General Electric's assurances, the Advisory Committee froze power levels for the Hanford reactors at their January 1958 levels until further studies were accomplished. The AEC quickly realized that prolonging this freeze on operating levels could lead to stalling any increases in production levels. Further, the economics of loading I&E slugs into the reactors would come into question, since the reactors would not be running at the expected higher powers. Though these factors did not immediately pose a problem, they could threaten future production levels. General Electric and the Production Division sought arguments to convince the ACRS that safety improvements made up for the perceived increased risk of a major accident.[26]

The Advisory Committee on Reactor Safeguards gradually came over to the side of power upgrades. One step towards this goal came in June 1958, when the committee agreed that previous power increases had not reduced the safety of the reactors. However, the committee resisted any further power increases until December 1958, when it was persuaded that plans for reactor confinement systems were being seriously considered by the Commission. Unintentional releases of fission products would be contained within the reactor building, reducing the risk to the outside environment.[27]

The Advisory Committee supported the installation of various filtration devices within the containment systems because they could block leakage of fission products to the surrounding area. From the standpoint of a complete failure of the primary coolant system, though, these modest *confinement* programs did not offer added protection. Instead, either a supplementary cooling system would need to be installed or a true *containment* vessel built, with each option costing several million dollars. Since the chance of such an accident remained "extremely remote" from GE's perspective, the expense of completely addressing such a catastrophe seemed excessive. General Electric continued design studies on such an alternative, if only to prod the Advisory Committee into approving further power upgrades.[28]

Caught in the tension between the Production Division, with its concern for quotas, and the safety committees, with their emphasis on the consequences of a catastrophe like that at Windscale, General Electric sought both technological and procedural solutions. The incremental modifications to the reactors emerging out of these managerial struggles accumulated into such significant total changes that one might say that all the reactors at Hanford were quite different machines at the end of 1958 than they had been when built, both in scale of production and in mechanics of operation. After the various upgrades at Hanford, General Electric ran each K reactor at 3,000 MW, C reactor at 1,600 MW, and the other piles at 1,350 MW. The contrast between design levels and power levels after the upgrading can be seen in Table 3.

Thaw in the Cold War

International events through the middle and late 1950s suggested to American leaders that the Cold War was very much alive, although a few developments suggested a lessening of tensions might be expected. On the one hand, the death of Joseph Stalin, the Korean truce in 1953, and the scheduling of "summit" talks between American and Soviet leaders provided signs that a thaw might come soon. On the other hand, the withdrawal of the French from Vietnam in 1954 in the face of Communist victories there, the Soviet suppression of the Hungarian uprising in October 1956, the successful Soviet orbiting of the Sputnik satellite in 1957, and the shooting down of an American U-2 spy plane over Russia in 1960 all suggested that the Soviet sphere of influence and Soviet technology would continue to threaten the West.

President Dwight Eisenhower's "Atoms for Peace" program, launched with fanfare in 1953, offered some hope that nuclear research, which had produced the threat of holocaust, might promise a more prosperous and peaceful world. That hope and promise would influence the shape and design of the last member of the production-reactor family, planned in the late 1950s and constructed in the early 1960s.

6 • Designing a Reactor for Peace and War

The building of the three new reactors at Hanford and the upgrading of both those and the older Hanford reactors brought plutonium production to the levels demanded by the Production Division. At the same time that production increased to meet the demands of the nuclear arms race, the Atomic Energy Commission began work on the "peaceful atom." With the Korean War truce, with President Eisenhower developing the Atoms for Peace plan, and with the 1954 Atomic Energy Act, the AEC began shifting resources to developing nuclear reactors for electrical power generation. The last production reactor built at Hanford represented an attempt to combine the mission of plutonium production with the mission of generation of electricity. That fourteenth and last reactor came into production in the 1960s, just as the earliest reactors reached old age and were ready for retirement. General Electric planned a modern reactor, to be safe, clean, and efficient for its dual purposes. As the company attempted to meet these policy goals, it chose particular technical options, giving the reactor a unique character.

A New Production Reactor

All eight earlier Hanford reactors used the once-through river water coolant system, following the original Du Pont design. The Jumbos had a supplementary ethylene glycol heat transfer system for heating the work spaces. Despite the incremental changes that had made them into more powerful machines with a host of different procedures, the first eight reactors quite clearly followed the conceptual design originally worked out by Greenewalt and Fermi under Groves's direction in 1943.

The new production reactor built at Hanford, eventually dubbed

N, was based on the old graphite-moderated design but had a different cooling system. A closed primary cooling loop of pressurized water ran through a heat-exchanger in a secondary loop. The primary cooling water for N reactor was pressurized so that it would remain liquid above 212°F, the secondary loop in the heat exchangers generated steam, the steam drove turbines, the turbines turned electrical generators, and the electricity thus generated was sold on the commercial power net to homes and industries in the Northwest. As a production reactor that could be converted to a power reactor, N was designated a "convertible" reactor in early discussions. Since GE did not work from existing blueprints originally drawn by Du Pont, as with earlier GE-built reactors at Hanford, GE engineers had the opportunity to build on their experience and to start afresh with N reactor.

Preliminary decisions made on paper regarding N began in 1957, with construction beginning in 1959, the power conversion features authorized in 1962, and the reactor completed in 1964. The reactor underwent an elaborate preplanned startup procedure in phases, through 1964; the generator was completed and operating in 1966. Despite the fact that only a few years separated the beginning of N's construction from the completion of the second round of Savannah River production reactors in 1955, the engineering approach at N was vastly different. Whereas the Savannah River reactors had been designed subsystem by subsystem, with many decisions postponed while others were implemented, N reactor at Hanford was completely designed before construction began. All the fundamental decisions, such as the overall power rating of the reactors, were postponed at Savannah River but firmed up for N reactor before construction began.

Several interacting factors accounted for the entirely different design procedure. For one thing, the policy decision on the unique function for N reactor specified that the reactor meet complex new technical requirements, optimizing between the needs of a production reactor and one designed to generate steam. In addition, the peacetime pace of the late 1950s and early 1960s allowed for more thoughtful planning than had been the case during the wartime urgency of the Korean War period. Third, the types of engineers doing the work at General Electric were quite different from those working for Du Pont, and systemwide planning came more naturally to them. Finally, the two companies brought quite different corporate cultures to the tasks.

Engineering and Corporate Styles

In 1943 there was no such field as nuclear engineering, and as we have seen, Du Pont's specialists in building factories for the production of chemicals brought their style to the wartime task. Chemical engineers were used to design plants in which the component subsystems could be changed as products were changed. These same engineers, some with experience gained in the Manhattan project, had built Savannah River, using their approach of flexibility. By contrast, N reactor at Hanford was designed by electrical engineers from GE. Electrical engineers then and later tended to think in a systemwide style.[1]

A switch toward the electrical engineering style at the Atomic Energy Commission was under way partly as a result of the growing influence of Hyman Rickover. Rickover, an electrical engineer by background and training, had led the project to design and build the successful nuclear propulsion reactor for submarines over the period 1949–55; and in 1953–57, he led the project to construct the first commercial power reactor at Shippingport, Pennsylvania, which followed a pressurized-water-cooled and -moderated design similar to that of reactors he planned for aircraft carrier propulsion. Rickover helped foster meticulous planning at General Electric when he obtained the services of GE's Knolls Laboratory in designing one of two submarine reactors in the early 1950s. Rickover's own technical staff worked closely with contractors, demanding close adherence to schedules, solutions to technical problems, and work to integrated systems as he built a network of suppliers, including General Electric.[2]

Rickover's influence over reactor issues would become even more profound in the next decade, but the change to a different style of design at Hanford suggests the nature of the "Rickover effect" upon the emerging field of nuclear engineering. One significant aspect of that effect was "systems" thinking. By contrast, the chemical engineering approach of Du Pont appeared almost haphazard, like the cut-and-try methods of craftsmen. In this regard, the change of engineering style for N reactor marked the evolution of the new profession of nuclear engineering from its World War II roots in chemical engineering to the electrical engineering–dominated style of the 1960s and later.[3]

The contrast between the corporate styles of Du Pont and GE, derived from the different business functions of the companies, also fostered the different approaches. Du Pont's primary business was constructing plants to produce various chemical products that would meet changing corporate policy in response to new developments and market demand. GE focused on manufacturing electrical equipment, from appliances through heavy industrial motors, transformers, switching equipment, and generators. With the Navy contracts at Knolls, GE moved into the research and design of reactors, not simply replicating earlier designs, as the company had with DR, H, and the slightly modified KE and KW reactors. At Knolls, company designers first worked on two sodium-cooled reactors, and then by 1956, still working for the Navy, they switched over to the design of light-water reactors.[4]

From Du Pont's point of view, an industrial plant should be flexibly designed to be able to manufacture a number of products; for General Electric, the plant—in this case the reactor—*was* the product.

Policy Choices Leading to N Reactor

N reactor's distinctive mission of convertibility from materials production to power production would require the long-drawn-out, careful planning process that characterized the GE approach. The AEC's choice of a convertible reactor had its background in a deep policy shift in the mid-1950s.

The policy emphasis on peaceful uses for atomic energy, long a concern of nuclear physicists, represented a dramatic change for the Commission. The world of atomic energy policy was very different in 1959 than it had been in 1950 or 1952. On 8 December 1953 President Eisenhower addressed the United Nations with his "Atoms for Peace" speech, which stated an American commitment to the development of peaceful uses for nuclear energy. Eisenhower held out the hope that nuclear energy could produce vast quantities of electricity and that the United States would take a central role in developing the technology. An international agency could be established to regulate the transfer of nuclear materials to fuel the new generation of power reactors. American industry would gain export markets for reactors and electrical generating equipment.

Reactions to Eisenhower's speech varied, but for the AEC it represented the inauguration of a new era. The 1954 Atomic Energy Act sought to implement Eisenhower's goals and to stimulate the development of nuclear reactors for the generation of electrical power and their eventual export to other nations, as well as to emphasize other peaceful uses for atomic energy. A relatively minor section of the 1954 act, section 44, authorized the Commission to sell electrical power generated in the course of weapons material production as by-product energy to public and private utilities or users. N reactor was planned to implement that section of the law.

Building N reactor with federal money, to generate electricity for commercial sale, raised in a slightly new form difficult political issues that had haunted the electrical industry since early in the twentieth century. In the 1920s, Congress had fought over the destiny of two federal power plants that had been built on the Tennessee River at Muscle Shoals, Alabama, during World War I, with the "public power" interests fighting against their sale to the private sector. In the early thirties the Tennessee Valley Authority (TVA) was created as a part of Roosevelt's New Deal to generate and market power in poverty-stricken southern Appalachia. Extending the principle to new areas, the federal government built dams and started power marketing administrations in Oklahoma and along the Columbia River in the Northwest to generate electricity and market it to municipalities and rural electric cooperatives. Utility companies resisted these developments as infringements by the government into an area of enterprise they believed more properly the province of the private sector. Since the days of the early New Deal, Democrats and a few Progressive Republicans had aligned on the public power side of this issue; conservative Republicans and some conservative Democrats were found on the private power side. As a federally funded and federally owned system selling energy in competition with private industry, N reactor recalled the debates and evoked much the same political array of support and opposition as those earlier federal hydroelectric projects.

Attempting to implement Eisenhower's goal of electrical power and the goals of the Atomic Energy Act of 1954 by developing specific technology, the Commission attempted a wide range of programs through the middle and late 1950s. The AEC and its friends on the Joint Committee worked in many ways to transfer the technology of nuclear reactors from the government to the private sector and to stimulate the necessary specific technology development required. The efforts yielded a lively competition among reactor conceptual designs, some of which proved valuable and others, unworkable. A five-year program announced in 1953 resulted in five projects by 1957, including the Shippingport reactor, which was built by Westinghouse under Rickover's supervision and was the first successful commercial large-scale power reactor in the United States. The effort expanded with nine more design projects in the Experimental Power Reactor Program, for the most part conducted at the Commission's national laboratories. Ten reactors were proposed under three rounds of another program, the Power Demonstration Reactor Program, first launched in January 1955. By June 1957, nine of the ten remained under study. These programs generated a va-

N reactor at Hanford (1967). Largest of the plutonium-producing reactors at this site, N also was designed to produce steam piped to a nearby electrical generating plant. The reactor is in the large building closest to the Columbia River in front of the tall smokestack. D and DR reactors can be seen in the distance. (U.S. Department of Energy)

riety of conceptual designs, with different moderator-coolant combinations. Only one, the ill-fated sodium-cooled Fermi reactor at Laguna Beach, Michigan, was in preliminary stages of construction by 1957.[5]

Over the course of 1958 and 1959, the Commission struggled to develop a comprehensive plan for future reactor development, inviting engineering and conceptual proposals for review. By February 1960, there were some 25 reactors in the AEC's long-range plan for power development, 5 of which were in operation or undergoing modification. General Electric, Westinghouse, and other manufacturers began to participate actively through the early 1960s in these projects. General Electric built the boiling-water design Big Rock Nuclear Power Plant for Consumers Power Company in Michigan under round 3 of the Power Demonstration Reactor Program, finishing it in 1962.[6]

Although these various power reactor efforts proceeded at the same time as the original planning for N reactor, N itself was not one of those activities, either from an organizational or a technical perspective. In its organizational context, N was funded, not by the section of the AEC devoted to power reactor development, but by the Production Division, the "weapons side" of the agency. Its design was classified and not transferred to the civilian sector. In fact, the only "transfer" was steam, piped through a fence surrounding the restricted Hanford area out to a generating plant constructed on the other side of the fence. The fence was physical, but it nicely symbolized the intellectual and ideational division between the weapons and the civilian sides and between the government preserve and the commercial world. Although N reactor met some of the broader contemporary goals of demonstrating that power could be a product of nuclear energy, the reactor was never presented as a part of one of AEC's various formal "demonstration" programs. However, in the atomic energy policy environment of the late 1950s, atomic energy for peaceful purposes was the stylish, *au courant* approach, the popular bandwagon. The Production Division was able to win some political allies in Congress and in the northwestern United States by hitching its new reactor to the contemporary drive for civilian applications, but that linkage was fraught with difficulties. For both technical and policy reasons, planners found it nearly impossible to design a reactor that would ideally fill both a plutonium production role and an electrical power generation role. Further, any effort to sell more government-produced power into the Northwest power market, which was already a battleground between private and public power interests, spelled further controversy.

In its organizational structure, the AEC maintained the separation of the defense and civilian sides, unaffected by some contemporary administrative reorganizations of the agency. At the end of the Eisenhower era, the Production Division reported to the assistant general manager for Manufacturing, while the Reactor Development Division, which handled the various demonstration programs, reported to the assistant general manager for Research and Industrial Development. Under the Kennedy Administration, the structure was changed with the addition of new assistant general managers and the proliferation of more planning offices. However, production reactors still remained under Manufacturing, power reactors under Research and Industrial Development.[7]

In 1950, when the first two Savannah River reactors had been authorized, the reactor population in the United States was small. Seven years later, when planning began on Hanford's N reactor, the various civilian development programs began to bear fruit. In 1950, in addition to the production reactors at Hanford and a few experimental reactors at the national laboratories, only the pressurized water reactor (Mark I) under Rickover and the BORAX boiling water test models were in development. By the time N reactor went critical more than a decade later, there were more than 35 major experimental and working power models and more under construction.[8]

Table 4 portrays the proliferation of types of reactors in the United States between the early 1950s and the early 1960s. The relative isolation of N reactor in the now-extended families of reactors, together with the competitive nature of power reactor development, is apparent. In this proliferation

Table 4. Summary of power and production reactors, 1951–1963, by type

Type/reactor	Designer	Total built, 1951–1963
Boiling water		11
BORAX-I–IV	Argonne	
EBWR	Argonne	
Vallecitos	GE	
BORAX-V	Argonne	
Dresden-1	GE	
Elk River	Allis-Chalmers	
Humboldt	GE	
Big Rock	GE	
Heavy-water-moderated and -cooled		8
HRE-2	Oak Ridge	
SRS production reactors R, P, L, K, C	Du Pont	
Heavy-water components test reactor	Du Pont	
Va.-Carolinas Heavy-Water Tube Reactor	Westinghouse	
Pressurized water		3
Shippingport	Westinghouse	
Yankee Rowe	Westinghouse	
Indian Point-1	Babcock & Wilcox	
Fast-breeder, sodium-cooled unmoderated[a]		3
EBR I	Argonne	
EBR II	Argonne	
Fermi	Atomic Power Development Association	

Table 4. *continued*

Type/reactor	Designer	Total built, 1951–1963
Graphite-moderated, once-through water-cooled		3
Hanford production reactors C, KE, KW	GE	
Experimental and other demonstration projects		8
HTRE series (3), INEL	GE	
Sodium reactor experiment	N. Am. Aviation	
Organic-moderated experimental	INEL	
Organic-cooled, Piqua	Atomics International	
Graphite-moderated, sodium-cooled, Hallam	Atomics International	
Graphite-moderated, pressurized-water cooled	GE	
Convertible production/power		1
Hanford N reactor	GE	

Note: [a] To produce reactor-grade plutonium and power.

Source: *Nuclear Reactors for Generating Electricity: U.S. Development from 1946 to 1963*, Rand Corporation R2116-NSF, June 1977; Jack Holl, Roger Anders, and Alice Buck, *The United States Civilian Nuclear Power Policy, 1954–1984: A Summary History* (Washington, D.C.: U.S. Department of Energy, 1986).

of types and designs, graphite-moderated reactors never became the model for power production in the United States, although the Soviet Union and Eastern Europe built dozens of RBMK power reactors with graphite moderation. Those reactors, however, used boiling water for cooling, rather than pressurized water as in N reactor. France and Britain used MAGNOX reactors—carbon-dioxide-cooled graphite reactors—through the 1950s and 1960s. For all these reasons, N reactor would live out its life in some isolation, as a reactor *sui generis;* it was the only one of the particular subspecies of pressurized-water-cooled graphite-moderated convertible reactors ever built.[9] An appreciation of just why N reactor took so long to design and how it became such a technological anomaly amidst the rapidly proliferating types and models requires an understanding of the unique mix of politics and policy that created it.

Origins of the Dual-Purpose Concept

Wilfrid E. Johnson, General Electric's general manager at Hanford, later took some pride in the fact that GE had explored the concept of a dual-purpose reactor at Hanford even before Eisenhower's Atoms for Peace speech and the revision of the Atomic Energy Act. As early as September 1952, General Electric undertook general exploratory work requested by the Atomic Energy Commission, and in 1953, the company had proposed a dual-purpose reactor for Hanford. Johnson noted that the "underlying thesis" of such a design was that "the transition from a wholly government-owned weapons-oriented enterprise to a private and public (non-federal) ownership of an electric power oriented industry could best be effected by a co-mingling of the economics." By 1955, however, the Commission had dropped the concept of "co-mingling" the weapons business and the power business because it was incompatible with the declared policy of separate development of a peaceful, exportable type of reactor. The president intended to put electricity in the hands of developing countries, but not machinery suitable for building nuclear weapons.[10]

According to Johnson, GE became interested in the possibility of reviving the dual-purpose concept "for technical reasons . . . having to do with containment of radioactive materials": the company had become interested in a reactor system that did not rely on a one-time pass-through of river water for cooling, but on a recirculating system. In that oblique fashion, Johnson referred to the fact that fractured fuel slugs at the earlier Hanford reactors had led to elevated levels of radioactivity in the river. With the higher operating levels, a certain level of slug failures and radioactive releases had become routine in the 1950s on the single-pass, once-through coolant system reactors at Hanford. Operating a recirculating coolant that did not escape to the river would require that it in turn be cooled by a secondary loop; that secondary loop presented the opportunity to generate steam for power. The planned system of heat exchangers between primary and secondary loop was far less risky to the river than the old systems because particles from slug failures stayed trapped in the contained primary loop.

However, a basic problem arose in using the "waste heat" from producing plutonium to generate electricity efficiently. The optimum operating temperature for a production reactor was below the boiling point because the hotter the water, the more frequent the failure of slugs. Yet the optimum operating temperature for heat transfer to a steam turbine system was well over boiling. Operation of the pressurized primary coolant water at about 250°F could accommodate both concerns, "optimizing" between two less-than-optimum choices to achieve the dual goals. As a production reactor, N could follow the traditional graphite model with channels for the fuel slugs. As a power reactor, its water coolant would be pressurized so it could go above 212°F without boiling, and it could transfer its heat to the secondary loop. Thus, optimizing between two essentially incompatible goals, a typical approach in many civil engineering projects, led to the unique conceptual design of graphite moderation, pressurized-water cooling.

In a letter of 4 May 1956, the AEC authorized General Electric to undertake production reactor studies on the production of electric power as a by-product, relying on section 44 of the 1954 Atomic Energy Act. After studies from 1956 to 1958, the JCAE recommended and Congress approved funding in fiscal year 1959 for the "convertible" reactor. Whether or not conversion itself would be economical, General Electric's Johnson admitted, was an "elusive" issue. "This question," he stated publicly when Congress considered appropriating funds to implement the conversion, "cannot be answered with any degree of finality because the answer depends on some very important and basic assumptions that can be made only by the government." For example, if one assumed that there was a good market for the power, one could conclude it was economical to convert. If one assumed there was no market, it was obviously not economical. Planners had to make arbitrary choices about other questions as well.[11]

Although the answers had to be assumptions, their consequences made the difference between a viable and efficient concept and a financial folly. What did plutonium really cost? What sort of revenues could be expected from the sale of electrical power? If electric power revenues came only in a later phase of the reactor's life, should the revenue from the later period be used to offset the cost of the earlier production of plutonium? Should the cost of converting the reactor from only plutonium and tritium production to power production be charged against the weapons phase, against the power phase, or against both? Should the cost of building in convertibility, estimated at $25 million, be considered as part of the plutonium-production phase, the power phase, or both? Would the power produced affect the local market and reduce the final price of the power marketed? Was there indeed even a market for the power in the hydropower-rich Northwest? If power were treated as a co-product with, rather than a by-product of, plutonium, then the cost of the reactor itself (not just the convertibility and conversion features) should be reflected in the price of the power. There were no "right" answers to these questions that could be determined in the abstract. Rather, each answer was a policy assumption. With so many imponderables, studies proliferated.

The answers derived from arbitrary assumptions; if one chose the most favorable set of assumptions, one came to positive conclusions about the practicality and value of a convertible reactor. If one assumed that hydropower could not supply the market, that electric power was a by-product of the reactor and its sale should defray the cost of the reactor, the plan was brilliant. However, equally arguable opposite assumptions proved convertibility a very poor concept and demonstrated that the new reactor should never be built. Some of the commentators, like Wilfrid Johnson, were astute in perceiving the arbitrary nature of the assumptions. Others simply made assumptions that fit their desired outcomes and plowed ahead with the arguments. The problem demonstrated nicely the difference between the nature of policy choices and the nature of engineering choices, and showed that the objectivity of engineering could not easily or simply be applied to reduce subjective policy choices to objective decisions.[12]

Johnson quite frankly told the Commission in the early stages of the planning that the same figures could be read in several ways. He was not ready to make all the necessary positive assumptions, and in 1958, he warned the AEC of potential difficulties in two very firm letters. Furthermore, General Electric was not anxious to operate a power plant, as it was not in "the power generating business, and under normal circumstances should not be expected to enter this field." The company built its equipment for sale to utilities and had no desire to become a utility itself. He later repeated similar objections directly to Senator Henry Jackson, Democrat from the state of Washington and a warm advocate for N reactor and for funding for Hanford more generally.[13]

Johnson pointed to other sorts of serious difficulties with convertibility from the beginning. Ordinary fossil and hydroelectric plants could successfully deal with the problems of variation in electrical network load. Shutting down a hydroelectric plant could conserve water behind a dam and shutting off a fossil fuel steam generator saved the fuel. However, efficient running of the nuclear materials production side of the operation required stable output, not variation up or down depending on network demand. If one entity operated the reactor for plutonium production and another operated the connected power plant, careful arrangements had to be made between the two organizations because they had to have opposite management goals. Johnson urged the Commission to work out all such considerations prior to committing to the convertible concept.[14]

Yet, for the AEC, the idea of convertibility held several attractions, and Johnson's caveats went unheeded. Some of the positive aspects, from the point of view of the AEC's Production Division, added up to good reasons to build the reactor. If the cost of the reactor could be partially offset by power sales, then the cost of producing plutonium could be reduced. Put another way, the cost of the reactor did not have to be represented as a complete expense to the weapons program but could be shown as a lesser amount. Furthermore, it made good economic sense to have a reactor capable of producing plutonium and tritium operating and at least partially paying for itself after the other plutonium-producing reactors were shut down.

If a production reactor could be used to produce electrical power, several politically symbolic and significant messages would be established. The plan won support in Congress as an attempt to demonstrate that the research and development that had gone into production reactors could finally pay off in a peaceful and benign fashion. Senator Jackson and other members of Congress, along with some pro-nuclear writers, viewed N reactor as part of the effort to put the United States in the forefront of benign uses of nuclear energy, quite in accord with Eisenhower's leadership, with the intent of the 1954 Atomic Energy Act, and with the AEC's attempt to adapt to those principles. Hanford, with its thousands of employees and its many businesses, could enter a transition into a peacetime economy. In spite of Wilfrid Johnson's hesitations, from the point of view of General Electric, then beginning to enter the business of building power reactors, the experience could further build its reputation. Even if the power reactor technology for N reactor itself remained classified, the identification of the firm and its engineers with the largest power-producing reactor in existence was good publicity. For nuclear power visionaries inside and outside the Commission who were skeptical about the slow rate of entry of private utilities into nuclear power, a successful AEC-funded convertible reactor might serve as a demonstration, as a competitive prod to get the private sector moving, and as a possible training ground for future power reactor personnel.

Behind the whole movement to find peaceful uses for atomic energy lay a deep psychological pressure. Guilt and horror at Hiroshima as the consequence of the triumph of science could be somewhat atoned by bringing cheap electricity from the atom. N reactor could play a part in that atonement, as could the whole commercial reactor program.[15]

Despite all such positive arguments, there were other ways in which N reactor would evoke widespread political enmity. As a government-funded and government-managed effort and as a government-owned reactor, N reactor fell squarely in the middle of the decades-old debates between public power advocates and the defenders of private power. In addition, nuclear power advocates themselves were divided between those who hoped to see development controlled by the private sector and those who hoped to keep the new source of electricity in government hands.

The interplay of some of these political overtones surrounding the planned reactor surfaced when the AEC asked the local utilities their positions regarding the proposed concept in 1959. Byron Price, chairman of a group of public and private utilities, the Pacific Northwest Utilities Conference Committee, some of whom later joined the consortium that operated the N-supplied generators and marketed the power, carefully studied the cluster of reports already developed by the AEC, Stone and Webster, Burns and Roe, General Electric, the Federal Power Commission, and the Bonneville Power Administration. Price's committee reported its observations back to the AEC, reflecting several cross-currents in its findings.[16]

Price reported that the Northwest could absorb the predicted power output, that the reactor could be built so as to bring on electrical generation in increments, and that the power output would have to be coordinated with the regional demands. Price noted that nuclear power should be competitive in price with hydro and steam power but that the estimated cost of the power from the reactor would be in excess of those costs. Some provision for supplemental power to serve as a reserve when the reactor was closed for maintenance would represent an additional cost. Although hydroelectric power was adequate to supply the needs of the region at the time, steam power might begin to supply future needs. When that happened, nuclear power would have to compete in price with steam power. Intangible benefits of the reactor included bringing research and development to the Northwest, including its local colleges, and serving as a training facility for future reactors. Weighing the positives and negatives, the committee endorsed the convertibility feature.[17]

Stronger advocates included congressional representatives of the region, particularly Senator Jackson. In general, in the 1950s Republicans lined up on the anti-public-power side, Democrats on the pro-public-power side of this issue, with a few local politicians of both parties generally supporting the expenditure of funds in Washington State. Friends of local utilities, however, tended to oppose any growth of the Bonneville Power Administration or of the federal power-producing sector more generally.[18]

In late 1959, the issue came to a head when Congress passed legislation authorizing $145 million for N reactor. The authorization specified the conceptual design of the reactor as a large-scale graphite-moderated, pressurized light-water-cooled reactor, with the cooling water carried to a heat exchanger system for driving steam turbines. The overall parameters described in the legislation were a 3,300-MW(t), 700-MW(e) power design rating, with "convertibility" but not "conversion" built in. Postponing the actual conversion costs to power production made it possible to gain the votes of some of the opponents of public power who supported the weapons program.

In advocating the new reactor, Senator Jackson recognized that up to another $100 million would be required to fund the final conversion. Nevertheless, he suggested that the United States was in competition with Britain and the Soviet Union in the nuclear field; he argued that to build the reactor was an act of patriotism. If the United States was to stay in the forefront of nuclear development, funding and building N reactor was crucial.[19]

Indeed, he had a point. The Soviets at Chelyabinsk in 1955 and the British at Calder Hall in 1956 had built production reactors that also generated electric power. France's G-1 dual-purpose reactor at Marcoule went into production in 1956. Two more French dual-purpose carbon dioxide gas–cooled reactors (G-2 and G-3) were under construction at the same location. General Electric, Westinghouse, General Atomics, and other American firms had started to market power reactor designs in Europe and Asia in competition with European firms, yet the USSR, Great Britain, and France were clearly ahead in the area of dual-purpose designs.[20]

Whether or not N reactor would eventually be worthwhile appeared to be a question that should be resolved by technical experts. Yet such expert opinion, when truly objective, yielded no simple answers. A March 1961 study by R. W. Beck, consulting utility engineers, commented on the effect of including the $25 million capital cost of the power convertibility of N reactor in the electric power costs. The study focused only on the $25 million that had been included as part of the original $145 million cost of the reactor; it would take at least another $95 million to actually effect the conversion. The study examined 18 combinations of different operating periods, different amortization periods, and different rates of interest in calculating the effect of the $25 million on the produced power costs. In a comment smacking of tautology, the report stated that a more limited set of variables could yield simpler results.[21]

In effect, the experts told the advocates that they could determine the outcome of their calculations by varying the assumptions and that there were no objective guidelines as to which assumptions were appropriate. Engineering and accounting, it appeared, did not serve well as tools for determining policy, but only for implementing and recording policy.

Other advocates, both those in favor of and those opposed to the project, turned to power policy experts, with similarly inconclusive results. Some experts appeared willing to make the necessary assumptions and produce positive or negative outcomes; others indicated that there were too many variables to properly evaluate the issues. For example, on 10 May 1961 Craig Hosmer, a Republican congressman from California and a member of the Joint Committee on Atomic Energy who vehemently opposed the idea of appropriating funds for conversion, reported on a poll that he took of 25 power experts employed by utility firms and universities. In general, the group reacted against spending the funds to develop N reactor as a power source. About two-thirds of his respondents saw "no substantial contribution to civilian technology," and about 85% believed power technology could be better advanced by spending the conversion costs on "a variety of other projects." However, Hosmer asked a very nonscientific set of leading questions: "1. Do you believe that conversion of the NPR to power production will make any significant contribution to the advancement of civilian power reactor technology in this country?" and "2. In your judgment, is the allocation of $95 million to conversion of the Hanford reactor the most fruitful investment that could be made in terms of developing peaceful uses of atomic energy?" His survey question went on to suggest at least four other possible uses for the funds as examples.[22]

Considering the phrasing of the Hosmer "poll," it was perhaps remarkable that several of the respondents approved of the contribution of N reactor to the knowledge of turbines, the knowledge of large-scale reactors, and experience with zirconium-aluminum ("zircalloy") tubes and fuel cladding, all of which would need further improvements with the increased power levels required for power generation. Some of the experts responding to Hosmer's questions complained that many of the reports on the features of N reactor were classified. Those with access could not comment for fear of divulging a classified point; those without access appeared bitter that the issue could not be intelligently reviewed by outsiders. At least one complained that no lessons could be learned because of "the way the government kept its books." Several commented that the issues involved economics or policy rather than technology.[23]

After a closely fought and bitter debate, a majority in Congress in 1961 expressed opposition to supporting federal funding for power generation, representing a temporary setback for the conversion idea at N reactor. The AEC then explored whether the newly formed power consortium Washington Public Power Supply System (WPPSS) might raise the funding through bond sales. These negotiations seemed, to public power opponents, intended to circumvent Congress. On 28 November 1961, however, WPPSS worked out an agreement with the AEC to buy power, funding the construction of the steam plant through bonds.[24]

In 1962, in response to congressional outcry from opponents, the JCAE, under Chet Holifield, stated that the Atomic Energy Act of 1954 authorized the AEC to market power from a production reactor and that it would not be necessary to go to Congress to ask for specific funding to build the power side. Funding had been approved to build the reactor, so no further votes were needed. Since funding had to come from the private sector to implement conversion, he argued, the Commission could go ahead. Holifield told his Republican colleagues that if they did not like the approach, they would have to amend the 1954 act to prevent the sale of power. Holifield and the Commission in effect told the Republicans in Congress that they had already lost the battle, despite the 1961 resolution against federal power sale.[25]

The apparently incompatible positions were reconciled with an essentially political compromise in the form of an opinion from President Kennedy's comptroller general, who stated that even if the conversion were privately funded, it would have to be authorized by Congress so as to ensure that the power sale strictly conformed to the provisions of the Atomic Energy Act. In this fashion, Congress would still play its authorizing role by having a chance to vote on the conversion.[26]

Safety and Design

At the same time that Congress debated the propriety of the N reactor idea, GE designers proceeded with the paper planning in great detail. In a review of safety and reliability issues at N reactor, AEC managers explicitly acknowledged the high degree of planning that went into the reactor, outlining a design process that derived details from preset criteria. GE first codified a series of safety requirements in a document issued in November 1957, and those requirements served as the design basis for N reactor. The criteria were formally expressed in detail under several headings: reactor coolant supply criteria, including primary, secondary, and last-ditch cooling systems; control criteria, specifically speed of control; and "total control," by which was meant a system that would allow for shutdown under any circumstances. That system arranged for thousands of marble-sized boron spheres to drop into the reactor; the boron balls would poison the reaction, without the disastrous effect of a liquid boron flood safety system. Liquid boron, while shutting down the reactor, would destroy it. The ball system had been developed in 1952–53 at Hanford and retrofitted to the early reactors there.

General Electric's methodical planning went from broad general technical criteria, to "scoping" of more detailed design criteria, which were approved at appropriate managerial levels. These design criteria then led to detailed designs, prototype procurement and testing, design modifications, and procurement of production equipment. The final product, GE executives claimed with some pride, "represents a second or third generation of design even though the engineered equipment (such as control-rod drives) may have been used for the first time at N-Reactor." The systems approach was in place; its virtue was that one could go through several iterations of progress on paper without spending money on steel and concrete. GE's design criteria required decisions to be made and revised and modernized before anything at all was built. When built, the company could claim, the resulting system would be both more modern and less expensive than reactors built by the older cut-and-try method.[27]

By showing how the safety requirements led to specific features, the AEC and GE both claimed that the N reactor design *derived from* safety considerations. To a great extent, the thorough planning and preapproval of designs before commitment to procurement did in fact derive from safety concerns about control, containment and emergency shutdown. General Electric spokesmen believed the reactor would rank with the best commercial reactors in its safety features, as they noted in a 1964 reactor safety report: "Although N-Reactor is not a typical power reactor, it should, logically, be no more subject to accident than a typical power reactor, and, in fact, has been designed to a reliability standard at least as stringent as those usual in power reactor design."[28] GE hoped that the probability of a major accident would be in the range of 1 in 100,000 to 1 in 1,000,000 for any year of reac-

N reactor's 44-foot-high charging face used a mechanical "arm" to fill more than 1,300 process tubes with fuel elements. (U.S. Department of Energy)

tor operation. Explaining one conceivable accident chain that could produce a major accident, the authors of the 1964 report demonstrated that such an event could arise only as a result of simultaneous accidents to three systems—the rod control system, the rod safety system, and the backup ball-drop system. Each system had to be safe to the point of 1:100 to reduce the likelihood of the triple accident to 1:1,000,000. The thinking was that

100 × 100 × 100 = 1,000,000 and that therefore one had to set an objective of an accident occurring less than once in 100 years for each subsystem. Testing could not achieve the standard, for it would require testing each system for 100 years; paradoxically, the system would wear out before the tests were completed. Instead, individual reliability estimates had to be generated for each subsystem; their combination into a total reliability estimate would establish the safety of the whole system. Spelled out in this logical but elementary fashion, the 2 July 1964 report represented an early use of a probabilistic method of describing risk.[29]

Indeed, the reactor was designed in a deductive way, from premises established by policymakers. In a series of "NPR System Parameters" documents, General Electric designers detailed the thermodynamic parameters of the reactor and primary loop, the physical dimensions of the components and equipment within the primary coolant system, the coolant properties of the reactor and the heat removal system, and the physical parameters that would affect the operation and the thermal power of the system. Unlike the case with the earlier production reactors, in which overall thermal output had been scaled up several times during design and early operation, with the N reactor the design remained firm, with the overall reactor rating set at about 4,000 MW(t). While actual startup would require stepping up to that design power level in several careful stages, the full design power level was established before construction and never increased after startup. The overall scale of the reactor was one of many features planned before construction; decisions on relatively minor considerations made by 1961 that remained fixed included the fuel element parameters, tube dimensions, predicted operating temperatures, coolant flow rates, and literally dozens of other specifications.[30]

At the AEC, well before final congressional approval for power operation of the reactor, representatives of GE met with the Production Division, the Reactor Operations Branch, and members of the Hanford Operations Office to discuss the plans for N reactor and raised safety and convertibility issues. By April 1962, the planners were able to review preliminary work on tentative values for reactor power level, number of loops, coolant temperatures, steam pressures, fuel exposure, reactivity coefficients, and fission product inventories; such calculations were in process a full 2 years before startup.[31]

In these early planning sessions, ACRS members raised questions, which

The N reactor control room and operations center. After the accident at Three Mile Island in 1979, Hanford built a simulated control room for use in its safety training program. (U.S. Department of Energy)

although not perceived as hostile or loaded, went to the heart of several of the N reactor issues. Such issues eventually would create problems for the reactor:

1. Was the confinement system adequate?
2. How would the reactor deal with external fluctuations in electrical power demand?
3. Would the power consortium, WPPSS, be a competent reactor operator?
4. Would N reactor be the only production reactor in operation when it was involved in power production?[32]

These questions remained unresolved before the reactor went into operation.

The elaborate planning system developed by General Electric, while time-consuming and preventing rapid progress from concept to construction to operation, had the virtue of producing an economical, safe, and modernized machine. But converting N reactor to electrical production as well as safe plutonium production required the jumping of a few more political hurdles.

Final Approvals

Under the AEC authorization act signed into law (Public Law 87–701) on 16 September 1962, following the comptroller general's resolution of the congressional log jam over the issue, the AEC had to make three "determinations" and submit the information to the JCAE before proceeding. The Joint Committee began to entertain those determinations on 27 September 1962. This resolution of the long-standing public-private power debate represented a compromise in which the Republican opponents of public power were allowed to save face. Congress would authorize Hanford generating facilities as long as the determinations that the operation would conform to the 1954 Atomic Energy Act were made.

Within 5 weeks, the comptroller general asserted that the AEC had provided the appropriate determinations for N reactor: (1) electric power would be a "by-product" of the reactor; (2) the sale of power could provide financial return to the U.S. treasury; and, (3) operation for power would enhance defense readiness. The AEC, the comptroller General ruled, had conformed to the original intent of the 1954 act by these determinations, and the Joint Committee simply endorsed the ruling.[33]

By November 1962, the AEC predicted cost overruns on the third reactor from the original $145 million to $205 million.[34] Nevertheless, Hanford defenders of the N reactor convertibility approach developed a number of presentations showing revenues for steam as a means of deferring the cost of plutonium and tritium. One such study concluded, "During the period of production need, this plant promises to be the Commission's most economic producer of nuclear defense materials." Despite earlier warnings that such calculations were dubious at best, defenders of the system found it difficult to resist the temptation to present them.[35]

The reactor construction and conversion project was completed on 15 April 1964, and at that time, General Electric anticipated that the reactor would be at full power by the fall of that year.[36] Preliminary runs at 10% of rated power were successful and the Advisory Committee on Reactor Safeguards approved stepwise upgrading to 75% of full power; in the meantime, the ACRS made only minor suggestions for improvement, modifications that did not affect the basic design. General Electric was to undertake the installation of filters in the ventilation system. The ACRS also recommended closer study of performance of various systems under unlikely but "maximum credible" accident scenarios.[37]

After the reactor went into operation in 1964, General Electric anticipated public questions about operational safety similar to those faced by the growing commercial reactor sector. When the company planned ahead with the AEC, looking to the day when the reactor would be converted to "Phase III—that is, the power-only, no-plutonium-production phase—it became apparent that despite all the safety design criteria, the reactor did not quite meet the standards being established for commercial reactors. Those variations from standards derived from its unique qualities. Despite the pressurized water system and the heat exchangers, the reactor would not meet commercial limits on radioactive releases to the river. In addition, the structure surrounding the reactor was built as a confinement system, to forestall emissions of radioactive gas or steam to the environment; this confinement system, while it could limit the effect of a catastrophic meltdown, did not meet commercial requirements for a full-scale containment system. In effect, N reactor might never be certifiable as meeting the emerging standards for water-borne and airborne radioactive emission as a power-only reactor, and thus it could ever move to Phase III.[38]

Production Reactors Up and Running

As built, N reactor's fuel slugs were 2.4 inches in diameter by 26 inches long, clad in zircalloy that was 0.03–0.04 inches thick. A partially automated system of fuel loading and unloading, hydraulic control rods, the boron-ball system for emergency scram, the air-filtered containment system, and a well-designed control room, all represented a modernized reactor that clearly was a product of 1960s technology, not 1940s technology.[39]

With the construction of N, the Atomic Energy Commission built the last of the genus, production reactors. There were two species in the genus. One was the nine-member graphite reactor group, all direct lineal descendants of CP-1, all built at Hanford; N reactor, with its pressurized water, modern features, and heat exchangers, was a specialized type of the graphite-moderated species reflecting the emerging new style of nuclear engineering. The second species was the group of five heavy-water reactors, all built at Savannah River, all modeled on Walter Zinn's conceptual design of CP-6 as modified and refined by the Du Pont engineers. By 1963–64, with all 14 running, reflecting all the power upgrades, the United States had over 36,000 MW(t) of production reactor power in operation, compared to the less than 750 MW(t) that had produced the strategic material for Trinity and Fat Man, dropped on Nagasaki. As a further measure of scale, in 1963, there was a total of about 860 MW(e) (or about 2,800 MW[t]) devoted to generating commercial electric power in the United States. When N reactor's power generators came on line in 1966 they approximately doubled the nation's total electrical output from reactors.[40]

The winds of change had started to blow, however, in the world of nuclear engineering, with President Eisenhower's emphasis on converting the atom to peaceful purposes and with the 1954 Atomic Energy Act. In the design and construction of N reactor, the AEC's Production Division and General Electric attempted to link materials production to the peacetime function of power generation. Further adaptation to change would become increasingly difficult over the next decade.

7 • Surviving Détente

The period from the late 1950s through the early 1970s was a time of difficult transition at both Savannah River and Hanford. At both sites the production reactors had been built and power levels increased under the urgent demands of wartime schedules, whether those of World War II, the early Cold War, or the intensified crisis atmosphere of the nuclear arms race. With the end of the Korean War in 1953 and the change of emphasis to peaceful use of atomic energy, the world of nuclear reactors had altered. Over the next decade and a half, decisions affecting reactor technology, reactor operation, and the very survival of the family of production reactors moved from behind closed doors into the open.

When the reactors had been built in World War II, the American public had no knowledge of their existence. Postwar upgrades and new reactors drew little notice. When new reactor construction sites were chosen, a few representatives of potential sites expressed interest, but the short flurry of public attention died out after Savannah River was selected in 1950. The AEC inherited from the Manhattan Engineer District both the formal and informal side of secrecy. On the formal side, the rules of classification, chainlink fences, an elaborate badging system, and controlled access to information and facilities all served to limit public awareness. On the informal side, the habit and tradition of providing no more information than was absolutely required meant that knowledge about production reactors remained limited. For such reasons, the myriad specific choices and broader policy decisions that shaped production reactor technology did not receive public exposure in the period 1942–52.

The Sixties: An Age of Transformations

Through the mid 1950s, President Eisenhower's commitment to converting reactor technology to peaceful purposes soon brought reactor concerns into a somewhat more public forum. With the construction of N reactor, quite suddenly some aspects of production reactor technology had become the center of overlapping political conflicts.

New developments in the early 1960s exposed further aspects of production reactor policy to the public. The original production reactors approached the end of their life spans. Safety evaluators within the AEC continued to have reservations about reactors that had been modified, repaired, rebuilt, and upgraded. With N reactor coming on-line and with the accumulation of plutonium nearing long-term weapons requirements, the need to keep alive the oldest reactors for the production of plutonium diminished. The Commission then had to face the issue of which ones to close first. The simplest but not necessarily the most scientific way to rank the reactors by risk was to regard the oldest as the most unsafe. Once reactors had been scheduled for closure, the Commission needed to deal with how to minimize the economic impact of the closures and to search for alternative ways to employ the facilities and the personnel to soften the impact. Thus, in the 1960s the dilemma of an excess production reactor population generated new issues about risk, closure scheduling, and diversification of the tasks at Hanford and Savannah River. Increasingly, such issues crept into a more open, and a more complex, forum. Decisions on each issue could no longer be made entirely behind the fences, insulated from public participation.

The sources of the transition from the closed-room style of internal decision-making to the more open process are diverse. From the end of the Eisenhower administration through the Kennedy and Johnson years, the United States underwent a deep cultural and political transformation. Values and ideas that had been part of the national consensus in the war and postwar years started to erode; old assumptions no longer seemed valid. In areas of life ranging from popular support for American foreign policy to trust in government and acceptance of corporate and academic leadership, profound changes were afoot. Some of the causes were international, such as the coming of age of a generation born since World War II; other causes seemed rooted in peculiarly American experiences, such as the discrediting of extreme anticommunism in the political career of Sen. Joseph McCarthy. Still other factors involved long-range trends, such as the proliferation of commercial television, the disruption of the traditional family, and deep-seated and unresolved social tensions over race and gender. The nuclear weapons complex, although insulated and isolated from the mainstream of American life by a curtain of classification, by a degree of geographic remoteness, and by the tradition of quiet decision making, could not be entirely immune to the underlying social changes of the 1960s. The transition to public involvement can be seen as an example of this broader transformation of American culture to a greater degree of participatory governance and as evidence of the declining popular trust in experts and scientists.

The specific case of transfer of production reactor decisions from government experts to a more public forum reflected unique and distinctive issues. Changing perceptions of risk and conflicts between national defense priorities and the economic interests of localities took the debate in particular, sometimes unpredictable directions. Actions by the Atomic Energy Commission itself contributed to the growing cynicism about government and distrust in decisions reached privately by leaders and experts. Some decision makers were shocked and irritated by the contrast between that new public cynicism and the euphoric patriotism of the early postwar years.

Bland statements from the AEC minimizing the risk of fallout from weapons testing inspired doubt rather than faith as contradictory evidence mounted. The exposure and consequent illness of Japanese fishermen aboard the fishing boat *Lucky Dragon* and of Marshall Islanders and American servicemen as a result of 1954 testing at Enewetak contributed to public doubts, as did congressional hearings on weapons fallout in 1957 and 1959. Hearings on AEC rule-making regarding the risk of reactor meltdown attracted some further public notice in 1959. Then, in the early 1960s, separate local grassroots groups objected to a proposed commercial reactor at Bodega Bay in California and to other reactors sited near Detroit and on Long Island, New York, because of the risks involved. In California, the opposition focused on the willingness of a public utility to despoil a scenic coastline and build near an earthquake fault. In Michigan and New York, the proximity of the planned reactors to population centers raised concerns about evacuation in the event of a catastrophe. Power companies preferred to site their reactors close to their heaviest consumer markets in metropolitan areas in order to minimize the loss of power over long-distance transmission lines. If the reactors were sited near cities, suburbanites feared their proximity; if sited remotely, nature-lovers raised an outcry. In these early and widely separated first protests lay hints and origins of what became a nationwide movement a decade later.[1]

In 1963–64, the AEC's assurances regarding the danger of radioactive exposure, reactor risk, and general issues of nuclear energy were no longer universally accepted, yet polls and votes on antinuclear propositions indicated that the majority of Americans still supported the concept of nuclear power development. The AEC tried to respond promptly to the new emphases on peace, safety, and public benefit, but the linkages between atomic energy and war, danger, and closely held secrets had been established. In the area of nuclear politics, as in other phases of American life, the decade following 1963 was one of difficult cultural adjustment.

At the heart of the adjustment was the issue of nuclear reactor risk. When and if a reactor had a major failure, its consequences could be disastrous. Even before the first Hanford production reactors had been built in 1943–44, the small knowledgeable circle of scientists and military men anticipated the potential dangers of reactors. It was for exactly that reason that one of the criteria for the selection of the site at Hanford had been geographic isolation. Even the Savannah River site, though in the heavily populated eastern United States, had been chosen in part for its relative isolation. But as the AEC sought to encourage power reactor development, a host of issues needed resolution, including how to compare the risk of reactors, how reactor risk increased as the reactors aged, and how various safety features should be evaluated. In the choice of alternate designs, some objective method of determining the safest system was desirable. The population of reactors was so small, and the consequences of an accident potentially so great, that it was impossible to apply the usual statistical measures of safety used on systems with more numerous individual cases of total failure, such as steam engines and automobiles.[2]

All of the production reactors except N reactor had been built before the movement to promote commercial reactors got under way. The early piles at Hanford had been shielded, but not "confined" or "contained," while the confinement structures at Savannah River would not meet commercial reactor standards. Protection against the dangers of meltdown derived from isolation; if a graphite reactor at Hanford caught fire, or a heavy-water reactor at Savannah River suffered a major pipe break, the intervening miles of uninhabited land between the site and the nearest residence offered some public protection. Yet winds and streams could carry radioactive smoke or liquid effluent for hundreds of miles, so the risk remained. As commercial reactors were planned for less isolated spots and as containment structures were designed to protect against public exposure, the dangers of the older reactors became more obvious by contrast. Public debates over the emerging regime of regulations and designs for commercial reactors that would minimize risk, reduce effluent, and protect worker safety became quite strident through the late 1960s and early 1970s. The older generation of production reactor cousins, ignored by the public when they had been created, simply did not conform to the evolving modern standards. If the AEC closed production reactors, it would resolve both the safety and the excess capacity issues. Yet reactor closure brought other public issues to the fore.

In this period, most of the people living near Savannah River and Hanford viewed closure of production reactors not as an environmental blessing but as a potential economic disaster. In both locations, the employment of thousands of technical workers had transformed the local economies. That effect was more pronounced in Washington State than in South Carolina, simply because the economic impact of Hanford was proportionally greater. Eastern Washington had been an arid, lightly settled region with an emerging agriculture based on irrigation; Hanford was the largest employer in the state for the whole region east of the Cascade Mountains. Furthermore, the three nearby towns in which Hanford workers lived all depended upon the government contract payroll. By 1964 there were nine production reactors at Hanford; those machines alone employed over 3,000 workers. At Savannah River, the population working at the site resided in a more dispersed fashion in surrounding counties, rather than being concentrated in specific neighboring communities. In both areas, however, representatives, senators, and state governors were sensitive to the impact of planning upon their constituents and worked to diversify employment opportunities.

By the time the AEC announced plans to close some of its production reactors, an intricate politics had already started to emerge in the related but distinct area of debates over commercial power reactors. Slowly, antinuclear sentiment became organized, at first in the widely dispersed local actions in California, Michigan, and New York. The issue of civilian power reactors may have served as a surrogate for opposition to or concerns about nuclear weapons. By opposing the vulnerable civilian reactors, concerned activists and members of the public more generally could express their hostility to nuclear weaponry. Eventually that organized hostility could affect the weapons programs. As the attacks against civilian power reactors mounted and as knowledge of the risks associated with them became more widespread, organizational hostility transferred to production reactors themselves.[3]

One of the sources of distrust of the Atomic Energy Commission was its own handling of the issue of radioactive fallout from atmospheric nuclear testing in the late 1950s and early 1960s. The Commission, with its tradition of expert guidance in decision making and controlled access to information, had no experience in or mechanisms for dealing frankly and openly with a potentially critical public. That institutional culture gradually spawned local, then national groups objecting to the risks at both the testing sites and at the reactor site at Hanford. The AEC's response of denial, dismissal of whistle-blowers, and resistance to outside investigation continued to exacerbate such fears and legitimate concerns over the 1960s and 1970s.[4]

Groups opposed to particular commercial reactors appealed for followers with the separate but related issues of risk to local real estate, concern for the natural environment, disarmament, nuclear safety, and participatory government. In the late 1960s, as utilities across the country placed orders for power reactors, concerned reactor neighbors in state after state took an interest in the issues. Gradually, the antinuclear movement coalesced.

Yet the engineers and industry spokesmen who sought to build more commercial nuclear power plants also expressed concern with international peace, with limiting reactor risk, and with controlling radioactive emissions. Pronuclear advocates claimed, with good evidence, that nuclear power was far cleaner than coal in the production of electricity and far safer to workers than the other energy systems based on coal or oil fuel. As the lines of argument and the advocates from the commercial reactor debate influenced production reactor issues, complexities and cross-currents abounded. For example, even advocates of various local nuclear reactor projects at Hanford and Savannah River, while very pronuclear in tone, grew suspicious of the AEC's closed decision-making process. Thus, pronuclear activists also began to demand more information as well as more participation in decisions. Risk, local economic impact, peaceful uses of the atom, and plutonium oversupply all became hot topics as production reactor debates moved into the open.

Land Use on Wahluke Slope

The low public perception of reactor risk in the 1950s is illustrated by a protest at Hanford over access to agricultural lands. During World War II, the Manhattan Engineer District had restricted access to land on the northern and eastern bank of the Columbia River, across from the Hanford reservation, in an area known as Wahluke Slope. This land provided a safety zone between the reactor areas and civilian populations. In response to postwar claims that the continued restriction of this land imposed "considerable hardship" on the landowners, in 1948 the AEC lifted some wartime restrictions, though it still did not permit extension to Wahluke Slope of any irrigation projects that the Department of the Interior sponsored elsewhere in the immediate area.[5]

As time passed without a major reactor accident, the Commission felt more confident about opening land on the Wahluke Slope to irrigation and development. In January 1952, under advice from the Reactor Safeguard Committee and a specially appointed Industrial Committee on Reactor Location Problems, the AEC released approximately 87,000 acres of a "secondary zone" on Wahluke Slope to irrigation works that included the Interior Department's Columbia River Basin project. At the same time, however, the Commission retained an area of the secondary zone that remained restricted from development, as well as a central zone directly across the river from the reactors to which all access was prohibited. As a precaution, the AEC directed its Hanford Operations Office to establish a warning system for slope residents in case of a reactor disaster. It also developed a public education campaign to inform residents of potential hazards associated with production reactors.[6]

Local residents remained unsatisfied with the slow release of land to farming and, during this period, unconcerned with reactor risk. Glenn Lee, editor of the area's *Tri-City Herald*, argued in May 1954 that the slope question was more an issue of the AEC's wish for the "power of a dictatorship" than the reality of a catastrophe at Hanford. Lee did not think the "thousands and thousands of beautiful acres" should remain "locked up" simply for use by the federal government. Eventually the AEC agreed, withdrawing its objection to irrigation development within the entire secondary zone of Wahluke Slope in December 1958 and allowing normal use to begin. The Commission justified its action on the basis that General Electric had installed more safety devices in the reactors, including confinement systems.[7] Glenn Lee's positions in this dispute show how in the particular situation at Hanford, one could be a local booster and comfortably pronuclear and at the same time also be pro-environmental, pro–participatory democracy, and anti-AEC.

These early debates over the use of Wahluke Slope reveal that in the 1950s, some residents near federal nuclear facilities feared government encroachment on their economic and property rights far more than they feared radioactivity. Even 10 years later, after reactor risk had become a politicized issue, it remained possible to find organized political groups at Hanford who regarded the loss of livelihood as a more dangerous prospect than radioactive hazards from reactors. To an extent, the early Wahluke Slope debates show very clearly how little the public near Hanford perceived nuclear facilities as risky up through 1959; indeed, the AEC was more concerned about such risks than the local activists thought appropriate. Such a contrast was not unusual, however, in isolated one-industry mining or timbering regions where local residents feared that zealous government inspectors might attack their central means of livelihood.[8]

As employees living downwind of the reactors began to associate frequent cancers and birth defects with their residence in the path of winds blowing past the reactors, they sometimes found themselves highly ambivalent about the conflict between their own loyalty to the nuclear establishment and to the national government which it served. Those institutions' apparent unconcern for public health and for their own personal distress was tough to take, difficult to understand. Several poignant and tragic stories of the "downwinders," victims of cancers and other diseases presumed to derive from radiation exposure, drew the attention of local and regional journalists.[9]

Reactor Risk: Theories and Practices

In the 1960s, when public concern about the risk of reactors increased, technical approaches to the study of such risk underwent a change. The older method, using "deterministic logic," served as the basis for design and early operation of the Hanford reactors and had characterized early safety planning at Savannah River. In this system, a possible worst-case accident was visualized, and then both design and operation were planned around methods of forestalling cause-event sequences that would lead to, or determine, such an outcome or accident. At Savannah River, a possible catastrophic accident received close attention through examination of various event sequences in worst-case scenarios. One Du Pont report studied the measures taken prior to Savannah River startup to prevent escape of radioactive materials to the environment. As part of the analysis of the risks involved, the reactors were described in great detail. The report spelled out the scram and control processes and reviewed a possible boiling accident, outlining a worst-case scenario.[10]

A 1953 Savannah River study focused on the consequences of the assumed failure of certain major systems—in this case, the failure of the six heavy-water recirculation pumps. This deterministic method—a "design basis accident approach"—did not consider unlikely combinations of unlikely minor problems that could combine in various ways into serious events; such extreme cases were not explored because they could be regarded as "noncredible." Instead, the 1953 report estimated the time taken by each step in one catastrophic accident that was presumed possible, showing that a scram had to be achieved in 81 seconds at the 300-MW level, and more quickly at higher power levels, to prevent the lithium-aluminum control rods from melting and producing a catastrophic accident.[11] Such calculations allowed designers to anticipate exactly how fast a scram had to take place at each level of power upgrade and to design accordingly. It was in just this fashion that deterministic risk analysis contributed to reactor design in the 1940s and 1950s.

By the 1960s, a new method of evaluating the risk of reactors began to emerge in the wider community of commercial power nuclear engineers. General Electric, with its growing experience in the power reactor manufacturing business, had more exposure to this new set of ideas than did the Du Pont operators at Savannah River. Some of the origins of the new method, called "probabilistic risk assessment," could be found in the academic education of the newly emerging profession of nuclear engineering. Ernst Frankel, who wrote a textbook for a course in systems reliability at the Massachusetts Institute of Technology (MIT), which he taught through the 1960s, drew on a number of works published in the period 1957–61 that gave mathematical models for assessing the reliability of complex systems and evaluating the probability of their failure.[12]

Frankel showed how, in the traditional deterministic approach to complex systems, engineers anticipated the physical causes of system failure and prevented them by redundancy (that is, duplication of crucial system elements), design, and maintenance. Under the probabilistic approach, engineers were able to estimate the numerical likelihood of failure of a system or subsystem with a given design and maintenance program. The two approaches were not incompatible even though their theoretical viewpoints were very different: with deterministic engineering, one looked to physical problems and their remedies; with probabilism, one assessed the system's reliability by examining the probability of failure of crucial components. The probabilistic approach was more likely to focus on the combined effect of multiple failures of subsystems. It was precisely this new method that General Electric engineers at Hanford employed in trying to convince the Advisory Commmittee on Reactor Safeguards (ACRS) that the risk of the combined failure of multiple subsystems at N reactor was in the one-in-one-million range. If three subsystems had to fail simultaneously, and the chance of each failing was less than one time in a hundred years of operation, the probability of all three at once was, for all practical purposes, nil. Douglas United Nuclear, which took over operation of N reactor in 1967, concurred that such a one-in-a-million, "postulated accident" could not happen.[13]

The different approaches, although theoretically complementary or compatible, could lead to somewhat different practical conclusions about how to improve the reliability of a system. Using determinism, a worst-case failure would be examined and the system designed to prevent the accident or to ameliorate its effects. Using probabilism, the reliability of a great variety of subsystems could be calculated and an overall, quantified judgment made about the total system; such an approach could generate a focus on design improvements on small but crucial parts of the total system. The increasing use of computers made easier the vast number of calculations required under the probabilistic approach. Furthermore, by calculating the minute likelihood of simultaneous or sequential failure of multiple safety systems, one might reasonably demonstrate that the likelihood of a meltdown accident was not believable—that it was "not credible."

In the late 1960s, several published papers brought probabilistic thinking more fully to the attention of nuclear engineers. One of these was a 1967 paper delivered at a Vienna conference of the International Atomic Energy Agency, F. R. Farmer's "Reactor Safety and Siting: A Proposed Risk Criterion." Farmer argued that risk could be measured by estimating the probability of a system's failure, and he made a distinction between acceptable and unacceptable risks.[14]

The growth of probabalistic risk assessment (PRA), as it came to be called in nuclear engineering, resembled in several respects the paradigm shift of a major scientific revolution. Some engineers who had been trained to use existing deterministic methods remained skeptical of the new system, warning particularly that it could lead to an unjustified reliance on numbers, many of which derived from guesses. The harshest critics thought it little more than an exercise in numerology. Practitioners of the new system showed that it addressed several difficulties not handled by the old, particularly the issue of combinations of minor causes into major consequences. That focus on the ability of the new system to deal with elements not adequately handled in the old system was similar in some ways to the pattern in scientific revolutions in which anomalies not explainable by the previous scientific theory could be explained by the new. As in better-known "paradigm shifts" or scientific revolutions, the development of probabilism in nuclear engineering was accompanied by heated controversy within the community of specialists.[15]

The contentious shift in risk analysis from determinism to probabilism proceeded among a rather restricted community of technical experts, including reactor designers and planners, mostly outside of any wide public notice. As such, it might be called a technological minirevolution, requiring a revolutionary change in viewpoint and practice among a very select group of technical experts. Although the minirevolution produced rethinking among such specialists, it was not the sort of world-shaking paradigm-shift associated with full-scale scientific revolutions. Nevertheless, the change in thinking about risk among the experts did have some long-range impacts on public perceptions, creating echoes in the press.[16]

PRA practitioners insisted that their method was supplementary to the earlier method; they saw no conflict with the earlier method, even though some engineers continued to resist probabilistic approaches as mere numbers games. It is reflective of the different engineering approaches at Hanford and Savannah River that Hanford's GE engineers, with their emphasis on systems planning, used early forms of PRA methods in planning N reactor in the 1960s, whereas Du Pont engineers at Savannah River did not take training in the newer methods and begin a probabilistic risk analysis of the Savannah reactors until the 1980s.

Although PRA emerged as a new way of evaluating the risk associated with nuclear reactors in the 1960s, concern for safety had always been an element in both design and operations. Du Pont was concerned about safety from its first days of running the Savannah River reactors, both in terms of industrial safety for workers and in terms of releases to the environment, and it applied its existing safety procedures. As a chemicals company working with highly dangerous substances, Du Pont had a long tradition of ensuring plant safety. Rather than assigning responsibility for safety to a separate office, Du Pont insisted that all line officers be personally responsible for the safety of the divisions under them as part of their central management mission. That corporate approach required considerable internal monitoring and reporting.[17]

In response to a request in 1962 by the AEC's local Savannah River Operations Office, Du Pont began submitting semiannual reports on incidents and a safety audit of performance at the Savannah River reactors. This documentation summarized problems every 6 months; earlier problems had been reported individually as reactor incidents (RIs). The semiannual reports to the Commission, focusing on trends, bore a more positive tone than the RIs, which had focused only on single incidents. In the semiannual reports, even a failure to reduce the number of incidents from one 6-month period to the next was presented as a measure of continued vigilance and a consistent level of performance in the face of changes and in view of the fact that the plant grew steadily older. Du Pont also addressed safety issues by improving reactor containment, power monitors, and internal radiation monitors.[18]

As the Commission's concern with reactor safety increased with the licensing and construction of less remote power reactors, operators at Savannah River examined even more closely some of the worst-case scenarios, continuing to use deterministic approaches. In 1965, in response to an expected ruling from the ACRS, the Atomic Energy Commission asked the contractors at both Hanford and Savannah River to begin planning to bring the production reactors into conformity with standards for commercial power reactors. Their responses to the Commission request showed the slight difference in how Du Pont's determinism and General Electric's probabilism worked out in practice.

The AEC evaluated the two sites in light of a Code of Federal Regulations rule regarding radiation protection standards and emissions (10 CFR 20) and another regarding reactor site criteria in case of meltdowns (10 CFR 100). Savannah River met the 10 CFR 20 standard on radioactive emissions to streams and the 10 CFR 100 site requirement regarding partial, but not extensive, meltdowns. In effect, Du Pont admitted that in the case of a worst-case accident, Savannah River did not meet the site standard. Hanford had only a narrow margin on release to streams; however, on 10 CFR 100, Hanford's explanation of the unlikely combination of events necessary to generate fuel melting made such an accident "incredible." By those probabilistic grounds, General Electric argued that Hanford met the 10 CFR 100 rule. Du Pont made no effort to argue that the worst case could not happen; General Electric could use its calculations to show that the worst case was beyond likelihood.[19]

In a 1967 safety report, a Du Pont safety officer at Savannah River asserted that the motivation for Du Pont to guard against a catastrophic accident was even higher than any measure of real public risk would suggest, because relatively minor effects could bring adverse publicity to the corporation. In an internal report, he drew particular attention to 20 incidents, criticizing, one by one, the assessment of the incidents as of minor significance. This unpublished report showed that some of Du Pont's own experts disagreed, sometimes heatedly, on how to assess, report, and respond to problems of control and radiation release. But even so, Du Pont personnel in this period regarded such issues as properly handled inside the company, seeking to avoid public misunderstanding or misapprehension.[20]

Through the 1950s and 1960s, Du Pont continued to operate the Savannah River reactors with a strict system of administrative controls that followed the contractor's own standard and methods for safety at its chemical plants. Du Pont's view was that a series of tried-and-true institutional mechanisms enforced safety: these included safety analysis reports, technical manuals, technical standards, mechanical standards, standard operating procedures, emergency procedures, test authorizations, reactor technology memoranda, facilities and equipment instructions, job plans, and maintenance procedures. For each of these institutional or procedural mechanisms, specific definitions, specific rules for issuance or modification, and specific responsibilities for implementation or authorization were all documented. This essentially Weberian bureaucratic method was detailed to the last step, reflecting Du Pont's institutional resistance to the revolution in risk assessment. The concept was that any alteration in procedures that might reduce safety would be thoroughly reviewed; mechanical or operational factors that might determine a bad outcome would be forestalled by good management.[21]

Both Du Pont and General Electric remained sensitive to charges that the reactors they operated were unsafe, either in terms of gradual radioactive or thermal pollution of streams or in terms of the risk of a major sudden incident or meltdown. But they were caught in a difficult position. In order to meet Production Division orders for material production, they had to keep the reactors running; all reactors involved risk, and most had some degree of accidental radioactive emission. Any explanations issued by the contractor of technical procedures designed to mitigate pollution or risk were difficult to express in classic public relations terms. Because such explanations were issued by the firm responsible for the equipment and its possible failure, almost any such statement appeared self-serving, especially if it was couched in general and nonspecific terms. On the other hand, detailed technical explanations could be so difficult for a layman to follow that they could have the unintended psychological effect of drawing attention to the risk itself, rather than to the complex methods used to reduce the risk. Yet, when the public or the ACRS expected explanations, the contractors had to make statements. Their position became increasingly difficult as public interest in the issues mounted.[22]

The problem of dealing with public perceptions of reactor risk had hardly existed in 1959; by the middle and late 1960s it had become an increasing administrative burden to both Du Pont and General Electric, as the issue moved into the open forum of the press and electronic media owing to contemporary expansion of commercial power reactors. Business leaders in the growing nuclear industry were troubled by exactly this "public perception" problem. At a panel presented in 1963 at the Atomic Industrial Forum, the new association of nucleonics firms, dominated by electrical engineers, several of the speakers noted the issue. C. Rogers McCullogh, former chairman of the ACRS and now vice president of a private nuclear firm, called it "a rising tide of criticism" of atomic energy. He felt the criticism was unfair and somehow politically motivated, since the nuclear industry was more concerned with safety and with explaining safety than were other industries.[23]

Peaceful Uses of Atomic Energy

Safety and risk were only one side of the public relations problem. With the growing emphasis on peaceful uses of the atom, and with the attempt through the presidencies of Lyndon Johnson and Richard Nixon to maintain détente, the production reactors and the whole nuclear weapons complex remained uncomfortable reminders of the fact that the United States was engaged in a nuclear arms race with the Soviet Union. Although the primary function of both Hanford and Savannah River remained the production of materials for nuclear weapons, Du Pont, for one, sought to characterize a variety of design changes as reflective of more peace-oriented research and development. This "R&D" emphasis, which included efforts to produce a number of experimental isotopes at Savannah River, was at first presented as evidence of the meeting of scientific challenges and of adapting to the new emphasis on peaceful uses of atomic energy. In 1964 AEC manager G. W. Bloch informed the Joint Committee on Atomic Energy that Savannah River was working on a plan to irradiate weapons-grade plutonium, transforming it by steps into americium and finally curium-244. Even though curium-244 was 300 times more toxic than plutonium-239, the Commission hoped to find a market for the isotope in Space Nuclear Auxiliary Power (SNAP) applications. The project was presented as evidence of the research and development capability at Savannah River.[24]

It soon became clear that the idea of simply operating the reactors, year in and year out, to produce the same weapons-related products without product improvement or an experimental program was difficult and uncomfortable for Du Pont and its technical staff. People with scientific training went to work for Du Pont, not to be machine operators, but to engage in research and development and to make "better things for better living," as the company proclaimed on the bottom of every piece of stationery. Operating a weapons-material production reactor on a routine basis, for quantity production, hardly met the official corporate ideal or the real professional needs of the employees, let alone the emerging ethos of Atoms for Peace, nor did it fit comfortably into the Du Pont culture, in which the pursuit of new products led to career advancement for executives. Du Pont sought visible and tangible connections between its weapons-material production efforts and peacetime applications. Production of curium-244 was only one such proposal among many to introduce variety and peaceful purposes into the Savannah River operation.[25]

With the growth of commercial reactors and with the AEC's continued emphasis on peaceful uses of the atom, Savannah River reflected the broader cultural shift. Difficult as it was, both the government and the contractor attempted to present production reactors and production facilities, all built for and dedicated to weapons manufacture, as somehow linked to, or convertible to, peaceful purposes. Efforts to produce isotopes, to harness the heavy-water design to power production, and to support power reactor work in other ways all became regular features of Du Pont and AEC public relations documents in the 1960s.

Du Pont participated from 1957 through 1962 in the commercial reactor development program, submitting a number of reports outlining how heavy-water reactors the company ran for weapons material production might be made into or designed for electrical power generation. The AEC requested that Du Pont prepare cost evaluations of heavy-water (HW) reactors at both the 500- and 1,000-MW scale, developing cost comparisons of heavy-water reactors with other types, including gas-cooled and light-water-cooled reactors. Du Pont did not wholeheartedly jump aboard the power reactor bandwagon, however. After thorough study of one type of heavy-water reactor, a boiling heavy-water-cooled pressure-tube reactor, Du Pont concluded that "large capacity reactors of this type are not competitive with conventional fossil fuel plants at this time in the USA."[26]

Central to the heavy-water power projects was the Heavy-Water Components Test Reactor (HWCTR), which had been proposed in 1956 to help commercial manufacturers evaluate various components to be used in possible heavy-water reactors built for generation of electricity. Over the period 1956–63, the test reactor was conceived, designed, and constructed by Du Pont at Savannah River at an approximate cost of $8 million. On this AEC-initiated project, Du Pont had neither the urgency that had characterized the building of the production reactors nor the free hand that it had enjoyed in building those reactors. Company officials suggested that the 7-year span from conception to operation did not meet their corporate standard for getting things built in a timely fashion. Du Pont remained uncomfortable with the total systems design style of reactor building so readily followed by General Electric. The HWCTR delays and the planning imposed by the commission simply did not match Du Pont's methods of plant design.[27]

About a year after completion of the HWCTR, the AEC informed the Joint Committee on Atomic Energy that the program had become a dead issue.[28] By the time of completion of the HWCTR, Westinghouse and General Electric had begun marketing of light-water reactors for power generation in earnest. The Commission's general manager explored, without success, whether NASA or a European agency would be interested in sharing continued operating expenses of the HWCTR. This short-lived program was an early example of the many setbacks and disappointments in the effort to develop peaceful programs at Hanford and Savannah River.

Nevertheless, the overall tone taken by the Savannah River operation in the early 1960s was quite upbeat, conveying an emphasis on the possible future conversion of Savannah River from a Cold War arsenal to a locale for civilian and peacetime research. The need to move to such a conversion soon became quite pressing.

Reactor Closings

As Du Pont and General Electric began addressing increased safety concerns in the 1960s, they faced a new challenge to their operations that spelled doom for members of the production reactor family and precipitated much greater public and political concern. Gordon Dean's 1952 prediction of a surplus of plutonium by the mid-1960s had been quite accurate. In fact, the intensive effort to increase production reactor power ratings in response to the per-

Table 5. Power level upgrades of production reactors

Reactor	Power level (MW[t])		
	1944–1950	1951–1955	1963–1964
Hanford B	1944: 250	1951: 435	1963: 1,940
Hanford D	1945: 250	1951: 435	1963: 2,005
Hanford F	1945: 250	1951: 435	1963: 1,935
Hanford DR	1949: 250	1951: 500+	1963: 1,925
Hanford H	1950: 250	1951: 500+	1963: 1,955
Hanford C		1952: 750	1963: 2,310
Savannah R		1953: 383	1963: 2,300–2,600
Savannah P		1954: 383	1963: 2,300–2,600
Savannah L		1954: 383	1963: 2,300–2,600
Savannah K		1954: 383	1963: 2,300–2,600
Savannah C		1955: 383	1963: 2,300–2,600
Hanford KW		1955: 1,800	1963: 4,400
Hanford KE		1955: 1,800	1963: 4,400
Hanford N			1964: 3,950
Total capacity (low estimate), 1964			36,300

Source: AEC 1140, "History of Expansion of AEC Production Facilities," pp. 32, 60–61, DOE Archives, RG 326, Secretariat, box 1435, folder I&P 14, History.

ceived immediate arms threat from the Soviet Union during the early 1950s advanced the date of plutonium surplus or glut. The weapons complex achieved a saturated market for plutonium by 1963, partly as a consequence of the steady upgrades in power over the preceding decade. Table 5 presents a cross section of the power ratings of the various production reactors up through the eve of the decision to close them.

In response to the plutonium supply levels, the AEC planned the retirement, first of the older plutonium-producing reactors at Hanford, with a cutback in enriched uranium production, and then of some of the dual-purpose, plutonium-tritium producers at Savannah River. As will be seen, the cutbacks were announced and carried out in a piecemeal fashion, with closure of a few major facilities at a time, over a period of 6 years. The Commission hoped to mitigate the economic impact by spacing out the shutdowns.[29] However, the Commission did not announce its policy openly. Rather, a few reactors would be closed with an announcement that production needs could be met with the remainder, and then, a year or two later, another round of closings would be announced, a process that only heightened, rather than quieted, the resultant political outcry.

On 8 January 1964, President Johnson announced the first reduction in plutonium and enriched uranium production in his State of the Union message.[30] The AEC scheduled four reactors for closure: F, DR, and H at Hanford, and R at Savannah River. These closings were scheduled through 1964 and 1965; the reactors were selected on the basis of being among the oldest and in the worst physical condition of all the members of the family.[31]

While the reason for closing the production reactors was that American plutonium requirements had been met, Johnson used the impending shutdowns as a gesture of international goodwill. On 21 January 1964, in his message to an 18-nation disarmament conference, Johnson announced that the United States was prepared to accept appropriate verification of scheduled reactor shutdowns, implicitly suggesting that the Soviet Union follow the American example.[32] The AEC then undertook the development of a verification system for international use to ensure that a production reactor had not been operated between verification visits. The closures, their relationship to disarmament, and the verification scheme were all given prominent mention in AEC publicity releases and in the Commission's annual reports.[33]

Commenting on the cutbacks in April 1964, President Johnson characterized them as reflecting "our desire to reduce tensions, and our unwillingness to risk weakness." Despite the effort to style the closures as a peace gesture, he also stated somewhat more frankly that he was "bringing production in line with need" and said that he anticipated that Soviet Premier Nikita Khrushchev would respond with similar cutbacks.[34] However, at no time in the public statements about the closing of plutonium production reactors did the AEC attempt to make clear that a permanent plutonium surplus had been achieved; placing some of the reactors on "standby" left a public impression that the closures might even be temporary. Eventually, when no "nonproduction" use for a reactor was found, its status would be altered to "permanent shutdown."[35]

Further closings in the mid-1960s continued to reflect the combined effects of oversupply, obsolescence, and détente. While the glut of plutonium and the obsolescence of the equipment made closures essential, the AEC continued to stress the implications for international peace: the closures were evidence of "restraint" in the weapons program and were "consistent with U.S. proposals in international disarmament discussions."[36]

Despite such high-minded implications, operators and local citizens at Savannah River and Hanford sought ways to keep their livelihoods. In 1964, regional power companies explored the idea of converting the Savannah

River reactors to generation of electricity. Private power companies lined up to support the idea, with 12 power companies writing to AEC chairman Glenn Seaborg that they would undertake and fund a study on converting a Savannah River reactor to power.[37] The group evaluated the possible conversion of R reactor at Savannah River to power use, outlining the specific modifications required. Their report concluded, however, that while such a conversion would be technically feasible, it would be too expensive.[38] The AEC operations office was apparently less than enthusiastic about the concept, for even before the power companies submitted their final report, it ordered R reactor closed. R was then cannibalized for parts useful for the surviving reactors at Savannah River.[39]

In another attempt to keep the reactors and the local economies running, Savannah River managers continued to argue for conversion of production reactors to the production of a variety of isotopes, claiming there was some need for them at NASA and in experimental science.[40] In 1966, the AEC acknowledged the idea in stilted language: "Previously determined reductions in weapons requirements have permitted the shutdown of four production reactors[,] and future requirements, while still uncertain, may permit utilization of production reactor capacity for non-weapons products."[41] The cold phrasing reflected the fact that isotope production at Savannah, while touted locally, received mixed reactions from headquarters.

As the planned shutdowns at Savannah River and Hanford became realities, members of Congress from South Carolina and Washington State sought ways of addressing the complaints about employee layoffs from their constituents.[42] In South Carolina, Gov. Donald Russell worked with community leaders and congressional delegations from both South Carolina and Georgia to lobby for further peaceful uses of the Savannah River reactors. Russell knew that South Carolina could not claim that it had a specially built town like Richland, Oak Ridge, or Los Alamos. Those towns might argue that the government that had created them owed them special consideration; such a situation did not prevail at Savannah River. Nevertheless, Governor Russell felt that fact did "not mitigate in any way similar problems in plutonium cutbacks at the Savannah River plant." The effect would be felt in "quite a number of communities and counties in South Carolina and Georgia, with an impact on the Southeast in general."[43]

In 1967, the AEC announced plans to close another Hanford reactor, stating that "currently projected requirements for national defense can be met with the reactors remaining in operation."[44] A year later, in January 1968,

Glenn Seaborg blandly explained that one more reactor at Savannah River and one more at Hanford would be shut down and placed on standby. Once again, a similar statement was given: "AEC review of currently projected requirements for reactor products in defense and civilian programs has indicated that the requirements can be met with fewer reactors than are now operating."[45] That decision reduced the total number to seven reactors: four at Hanford, three at Savannah River. Each statement could be read to mean that the announced closures were final. When followed shortly by yet another announcement, the cumulative effect increasingly frustrated the politicians who tried to satisfy voters in the affected areas. Through this period, local politicians and their constituents overwhelmingly resisted closures due to the prospect of lost employment. Despite growing complaints of "downwinders" about health hazards, those closest to the sites tended to be dependent on employment and most pro-reactor.

The AEC did not close reactors on the basis of the oldest first, or even on the basis of selecting the ones with the most physical problems. In 1968, for example, the Commission selected Savannah River's L reactor for closure, over the more troublesome C reactor. Although L was slightly older than C, C had developed a history of minute heavy water-leaks, adding up to 50 to 100 liters per day. In support of the choice of L for closing, R. E. Hollingsworth of the AEC explained to an increasingly skeptical Joint Committee that despite the history of leakage at C—leaks now assumed to be "dormant"—other factors required that L rather than C be chosen for shutdown, including the relative production efficiencies of the two reactors when it came to tritium and a cost comparison of the maintenance and reconditioning requirements of the two reactors.[46] C reactor continued to leak until its closure in 1987. In effect, it was cheaper to keep Savannah River C reactor in production of tritium than to reconfigure L reactor for that product alone.

The AEC surprised its political allies further in 1969 when, under pressure to cut the budget, it announced still another group of closures at Hanford. As the closures and the layoffs continued, the local community became a bit jaded about reassurances from Washington. Local representatives, with newspaperman Glenn Lee as spokesman, complained bitterly to Commission chairman Glenn Seaborg in January 1969. The announcement of the closure of more reactors at Hanford had "shaken the community more severely, and done more psychological damage than anything which has happened in the last five years' time." Both Senators Warren Magnuson and Henry Jackson had been assured by the AEC within the previous month that there were to be no more cutbacks; Lee and the senators regarded the planned cutback as a betrayal.[47] With no notice or opportunity for reconsideration, the reactors built in the 1950s—Hanford's C, KE, and KW—would be shut down.

By 1971, the round of closures was complete. Only the youngest four of the fourteen production reactors remained in operation: P, K, and C at Savannah River, and N reactor at Hanford. The impact at Hanford was, of course, more profound: eight out of the nine reactors had closed there, whereas at Savannah River, two out of five had closed. Furthermore, as one of the MED-built communities, Richland, Washington, had no other raison d'être besides its nuclear work; with the closing of the reactors, Richland citizens easily visualized a return to 1940, when, they imagined, their community had been a dusty haunt of rabbits, tumbleweed, and an occasional buzzard. In reality, in 1940, developing orchards and vineyards had already begun to benefit from irrigation and cheap electric power before the MED and Du Pont arrived. But the town dwellers, dependent on government and contractor employment, feared that further closures would reduce Richland to a ghost town.

The drawn-out process of closing had reduced the number of layoffs at any one time. Yet the AEC's oft-repeated announcements that the current reactors were an appropriate number, followed by repeated further shutdowns, soured relations between the AEC managers on the one hand and contractors, employees, local community leaders, and congressional representatives on the other. At no time in the process did the AEC explicitly state that the nation had more plutonium than required but only that remaining reactors could meet requirements. Under the culture and practice of secrecy, more detail on the nuclear stockpile was not considered public information. Rather than minimizing public and political impact, the 8-year round of closures had maximized it, with a new blow to the Hanford region economy almost every year, as shown in Table 6.

Production reactor shutdowns in the 1960s represented more than the AEC's efforts to eliminate surplus, expensive, unneeded, and aging equipment. With the closures, the total thermal megawattage of production reactors declined, while the output of the electrical-generating commercial reactors steadily climbed.

The transition from reactors for weapons to reactors for peace at first was a matter of rhetoric and wishful thinking. However, one very tangible measure of the rate of implementation of that new emphasis in the 1960s

Table 6. Reactor closings during the plutonium glut period, 1964–1971

Reactor	Year closed	Approximate years operated
Savannah R	1964	11
Hanford DR	1964	14
Hanford F	1965	20
Hanford H	1965	16
Hanford D	1967	22
Savannah L	1968	14[a]
Hanford B	1968	22[b]
Hanford C	1969	17
Hanford KW	1970	15
Hanford KE	1971	16
Average approximate years of operation		17

Note: Totals are approximate in that reactor shutdowns for refueling and maintenance are not deducted; years have been rounded off to nearest full year.

Reactors still in operation in 1972 were P, K, and C at Savannah River and N at Hanford.

[a] L was later reopened between 1985 and 1988.

[b] B was was closed temporarily in 1946–47.

Source: AEC annual reports; EGG Box 2, 5661.1.7.2 Background Briefing.

was the strictly objective number representing the total thermal megawattage of production reactors in the United States compared to the total thermal megawattage of the new generation of power reactors. The shrinkage of the production reactor family and the growth of the electrical power reactor family show exactly the rate and timing of the actual supplanting of the weapons-related use of reactors by the peaceful use of reactors. Five large commercial power reactors came on line in 1970, while C and KW at Hanford closed in 1969–70, tipping the balance.[48] In the United States by the end of 1970, the total power of reactors devoted to electrical generation exceeded that of the production reactors (Fig. 2). As a concrete measure of the peaceful atom compared to the weapons atom, megawattage told the story of the 1960s transition.

Local Pressures at Hanford

For the men and women employed at the production reactors, the end of an era was not seen in statistical terms nor in terms of a gradual changeover

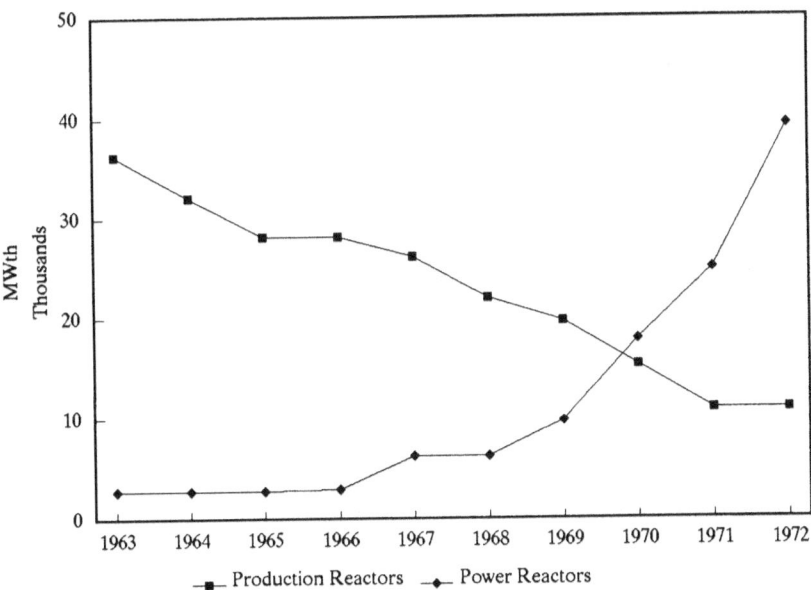

Fig. 2. Reactors for war and peace, 1963–1976. Before the mid-1960s, most American nuclear reactor energy was devoted to making weapons-grade strategic materials in production reactors. By 1970, the growing number of commercial power reactors represented the largest percentage of reactor megawattage, documenting the shift to "Atoms for Peace." (Figures from Jack Holl, Roger Anders, and Alice Buck, *The United States Civilian Nuclear Power Policy, 1954–1984: A Summary History* [Washington, D.C.: DOE, 1986], and AEC 1140, "History of Expansion of AEC Production Facilities," DOE Archives, RG 326, Secretariat, box 1435, folder I&P 14, History.)

from weaponry to peacetime hardware. Rather, it was a matter of income, employment and for those at Hanford, the very existence of a community. Closing the eight Hanford production reactors had the effect of moving the strictly technical issues of production reactors not only into the political realm of Congress and its committees but into the grassroots realm of community action and community politics. The AEC's actions in spreading the shutdowns out over nearly a decade had the effect of slowly building a local pronuclear constituency that became organized and more vocal. Although in favor of nuclear research and development, that constituency, like Glenn Lee, remained suspicious of decisions reached by the AEC or its successors in Washington. The closures created groups that tried to bring a popular form of participatory governance to the formerly closed decision-making process of the AEC. The groups and alliances continued to attempt to shape reactor technical choices over the following decades. The origins of the grassroots alliances lay in the AEC's own policies. Paradoxically, the organizations could be simultaneously supportive of nuclear research and nuclear industry, and critical of or even hostile to the Atomic Energy Commission.

On 8 November 1962, the Atomic Energy Commission announced a policy of cooperation in industrial development efforts with the community of Richland at Hanford to ease the transition to peaceful uses of the site. Attempting to show sensitivity to local feelings, the AEC stated that "such cooperation will not be a substitute for community leadership and initiative, nor intrude in the management of local affairs by the elected representatives of the people." In particular, the plan stressed the use of government-owned land for industrial use, the funneling of non-AEC government work to the area, planning for industrial development, and funding for educational and tourism activities.[49]

The AEC's cooperation policy took the form of the so-called Slaton Report, which offered assistance to both Richland and Oak Ridge, two government-built communities facing transitions. In the report the AEC promised cooperation but urged local initiative.[50] Based on this encouragement, in January 1963, a group of local businessmen in the Hanford area formed the Tri-City Nuclear Industrial Council. The tri-city area, with a combined population of about 55,000, included the nearby communities of Pasco and Kennewick, as well as the government-built community of Richland, which directly bordered the Hanford Reservation on the downriver, southern side. Richland had just been incorporated as a community in 1958.[51]

The local press became involved, working with local bankers and businessmen to try to stimulate interest among industrialists in leasing facilities at the Hanford site and making use of the local technically trained labor pool. The efforts, while smacking of local boosterism and often colored by a self-delusional mix of optimism and jawboning, did eventually result in additions to the mission at Hanford and the construction of both experimental and electrical generation reactors there.[52]

Newspaperman Glenn Lee became a prominent spokesman for the group. The activist editor of the *Tri-City Herald* addressed a group in Seattle in February 1964 to "give them the story behind the headlines." Lee indicated that the planned closure of three reactors and plans for General Electric to pull out as chief operating contractor led many people to believe that Richland would be "boarded up" and turned into a "ghost town." He bitterly claimed

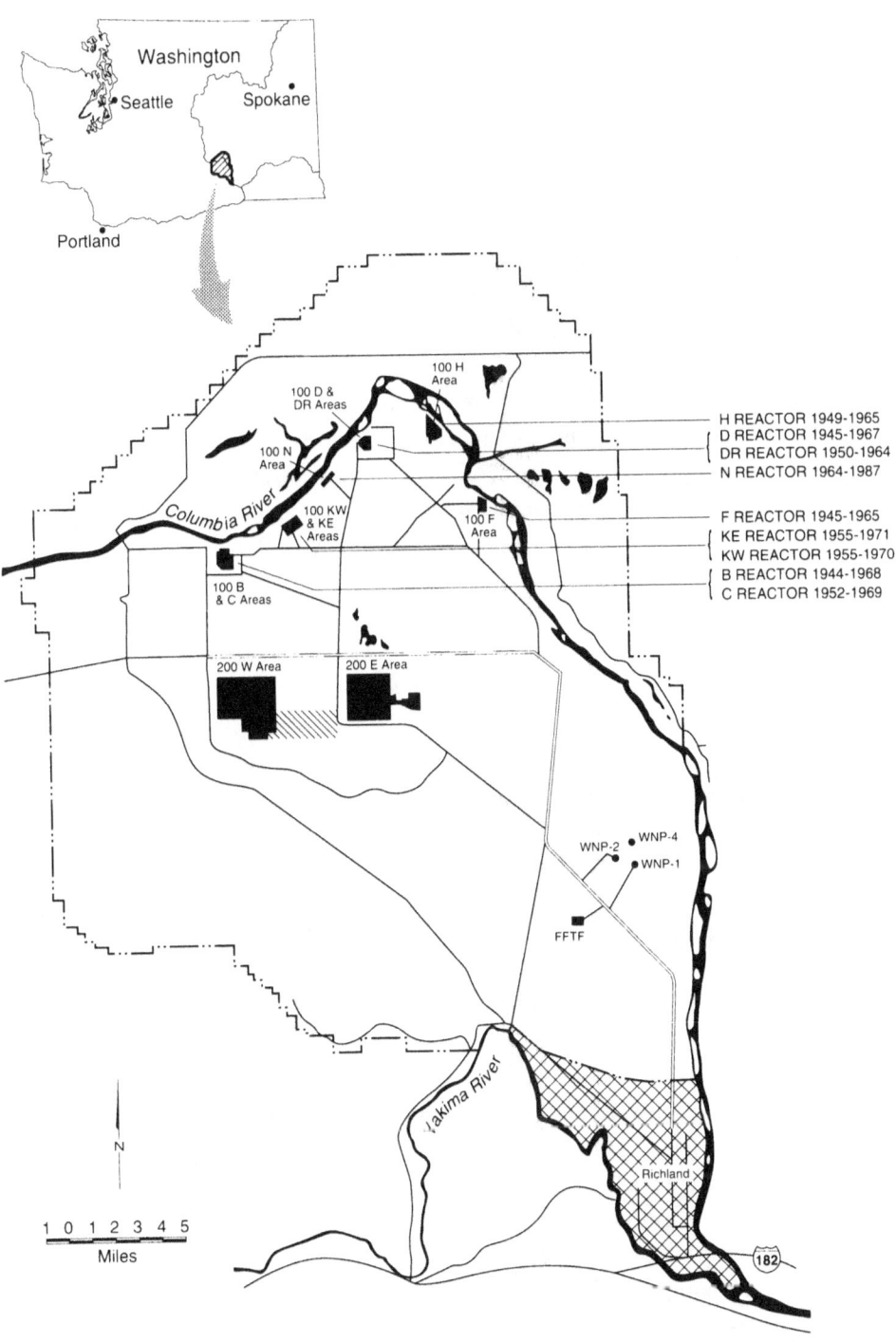

The Hanford site.

that although the Hanford area benefited from the $2 billion invested there, the area had been "a slave and a captive of the plutonium production department" of the AEC. He decried the Commission's "hammerlock on this plant, its secrets, its people." Lee credited the local community with foreseeing the negative effects of dependency on the AEC and indicated that local support for the Washington Public Power Supply System (WPPSS) generating system at N reactor had stimulated that development. He saw N-reactor commercial power as the first nonfederal "proposition" behind "the plutonium curtain." He continued to be critical of the Commission and its closed-door methods while remaining a nuclear enthusiast, a position not unlike that of many of his local newspaper readers.[53]

Working through the Tri-City Nuclear Industrial Council, Lee as secretary kept in touch with the JCAE, advocating a variety of specific projects for Hanford. In 1964 he pushed for a fast-fuel test reactor, for a fuel reprocessing center, and for unspecified work for NASA.[54]

Diversification Efforts

With promoters like Glenn Lee keeping the fate of Hanford in the public eye, the AEC worked to implement diversification to solve the area's employment problems. As the term came to be employed at Hanford, "diversification" had two meanings. Both the employers and the products would be diversified. First, a number of separate contractors would replace General Electric, which phased out its participation at Hanford in the mid-1960s. Separate aspects of the Hanford operation were assigned to different contractors, with the main N reactor operation transferred first to Douglas United Nuclear and later to Westinghouse. Glenn Seaborg told Senator Pastore of the JCAE that multiple contractors were "in the best interests" of both GE and the government, would help in "stimulating commercial diversification," and would "contribute to the future development of the communities in the Hanford area."[55]

The rationale was that Richland would no longer be dependent on a single firm for employment. Furthermore, the various firms would attempt to attract private business of one kind or another to their separate functions, in physics, engineering, chemistry, and computer work. Both the AEC and Lee's group visualized a pattern in which Richland would be converted from a GOCO company town to an industrial community with a diverse corporate base.

The Atomic Energy Commission intended this diversification of contractor organizations to lead to the second meaning of the term—diversification away from AEC contract work to a more varied range of products and clients for its industries. Early in the planning, the Tri-City Nuclear Industrial Council supported this second form of diversification by contacting NASA and the Department of Defense in hopes of lining up plant or laboratory development at Hanford. An AEC–General Electric study group chartered to look into alternate sources of employment tried to formulate a Hanford diversification program. The study group issued several reports. One studied fuel fabrication capabilities, a second evaluated Hanford as a site for experimental reactors, and a third report in the form of a brochure described Hanford's capabilities. The reports combined technical information with evaluations of facility capabilities and personnel qualifications and could provide information for prospective industrial investors. The switch away from production reactors could represent a new era in nuclear matters, and Hanford technical experts made explicit the effort to adapt to the changed times. A fourth report, "The Potential for Diversification of the Hanford Area and the Tri-Cities," issued in January 1964, concluded that the Hanford area was too dependent on AEC-funded plutonium production. The report suggested that at least 2,000 jobs would be lost over the period 1964–1968 with the reactor closings and that it was unlikely that government employment would replace the jobs. An "aggressive community effort" was required to make up the loss.[56]

This report also explored means of making use of one or more of the closed production reactors to produce either uranium-233 or polonium-210, or, after conversion, to generate power.[57] Other suggestions included using one of the older reactors as a test reactor or as a training unit for reactor operators. The study group found all such proposals unworkable, for the simple reason that the closed reactors were of unique design. Their light-water cooling and graphite moderation were not suited to power production. For the same reason, they did not represent either a good basis for alternate isotope production or for training. Training on a once-through water-cooled graphite-moderated reactor had little bearing on the operational needs of the new pressurized water and boiling water reactors that were emerging as the standards for American commercial power production through the 1960s. Everything differed, from the basic physics through the safety systems and the instrumentation in the control room. However, since *each* of the closed reactors represented as many as 400 jobs, the study group strongly urged further consideration of alternate reactor mission possibilities.

The Commission worked to attract vigorous institutions to Hanford to engage in new projects. Battelle Memorial Institute agreed to operate the Hanford Laboratories, which employed 1,842 people, beginning in January 1965. Battelle announced its intention to invest $5 million to attempt to attract more private work to the laboratory. Commission chairman Glenn Seaborg publicized the successful wooing of Battelle at the 1964 Seattle World's Fair when he spoke at the Hanford Exhibit. The AEC encouraged other initiatives by research and development groups, offering generous lease terms to land on the Hanford reservation, including 85 acres to the University of Washington and 1,000 acres to the state of Washington for nuclear industrial development. Another 400 acres were sold outright to the city of Richland.[58]

In a glowing letter of praise, President Lyndon Johnson suggested that the diversification plans had his blessing. He wrote to Glenn Seaborg early in 1965, lauding the Commission for advanced planning done without much fanfare. He credited Seaborg with foresight: "The cutbacks in special nuclear materials production were planned sufficiently in advance so that the Commission, in cooperation with the local officials and business and labor people, could take appropriate actions, such as diversification programs, to minimize any significant economic impacts."[59] It was true that the AEC had planned in advance, but Johnson was a little premature in his praise, as the process was just beginning.

Some experts were less sanguine. The Advisory Committee for Reactor Safeguards, still representing a somewhat independent voice, met with the Production Division at the beginning of the diversification program and issued a four-point critique of the effort. At the heart of their concerns was the issue of how to fit obsolete equipment into the emerging safety requirements of the 1960s. The new efforts created a backlog of work for the ACRS that could endanger safety if it created pressures for premature approval of new devices or programs. The Savannah high flux reactor proposal, the curium-244 loading, and the U-233 programs all came at once.[60]

The ACRS asked what safety standards should apply when production reactors were converted from making plutonium and tritium to production of alternate isotopes. Since plutonium and tritium were strategic materials for weapons, national security reasons might have justified a degree of risk in the design and operation of the reactors. However, production of peaceful-use isotopes like curium, which had no such defense-related justification, required a higher and more restrictive set of standards, such as those in the Code of Federal Regulations for private power and private isotope-producing reactors. If a reactor were converted to peacetime use, it would have to meet peacetime standards, and the production reactors simply did not do so. The ACRS noted that the Atomic Energy Commission provided "no real answer" to this objection. In addition, the diversification at Hanford, which resulted in a variety of contractors, would lead to a dispersal of authority and, hence, a diminution of safety responsibility unless a coordinated safety plan was developed and implemented. The AEC promised such a plan.[61]

Even if these new programs met safety concerns, the ACRS had further misgivings about diversification. The committee wondered if there was a viable market for the new isotopes, whose production had been used to justify the entire diversification process in the first place. The ACRS also thought that Hanford and Savannah River might compete for the limited number of viable alternative programs. This scramble for work, in the opinion of the ACRS, could have an "adverse effect on safety."[62]

One bright note for the Hanford communities through the period of closures was the operation of N reactor and its steam plant. Although a labor dispute closed the facility in September 1967 for several months, operation of the reactor in 1966 with its electrical generation system had proven quite successful. WPPSS, the consortium of public utility districts that operated the generating plant, announced that low construction costs and operating revenues allowed immediate retirement of about $25 million of the $122 million in bonds that had been raised to finance the project. Further, while the reactor-generating system operated through 1966, it produced 35% of the nuclear-generated electric power on line in the whole nation at the time.[63]

By the end of the production reactor closure period, Hanford's vigorous diversification effort began to show signs of paying off. In 1970, the Richland City Council urged the creation of a nuclear industrial park on the Hanford site, at which a series of commercial reactors would be constructed.[64] Although the federal reservation was never designated as such a Park, Hanford did become the site of several commercial and test reactors over the 1970s, which met some of the objectives of the local groups.

The Fast-Fuel Test Reactor (FFTR), which took nearly a decade to bring to fruition, was one such effort. The name was later changed to the Fast Flux Test Reactor and then to the Fast Flux Test Facility (FFTF). The FFTR under any of its names was a 400-MW sodium-cooled fast reactor that could run on and test either a plutonium oxide or uranium oxide fuel, or various mixes of the two, in stainless steel ceramic-metallic units called "cermets." As the AEC contemplated a breeder reactor program that would use reactors to produce, not weapons material, but reactor-grade plutonium fuels for further power generation, the test reactor would be a key research instrument, testing the fuels themselves in high-neutron-flux conditions. Later, with the termination of the breeder program under President Jimmy Carter, the mission of the reactor had to be altered to ensure its survival. The FFTR was completed in 1975 and went into operation as a research and testing facility for the AEC and its successor agencies.[65]

Other long-range successful diversification efforts in the early 1970s included the construction of reactors for electrical generation. WPPSS built Washington Nuclear Power (WNP) number 2 at 1100 MW(e) and WNP-4 at 1220 MW(e). In addition, the consortium planned but only partially completed WNP-1, also at the 1220-MW(e) scale.[66] A series of commercial light-water reactors, these large WNP reactors provided both employment for Richland residents and a partial raison d'être for Hanford, much as proposed by the Richland City Council in 1970.

Monopoly and Overproduction

The manner in which the Atomic Energy Commission dealt with the aging reactors and the plutonium oversupply of the 1960s led to several consequences that affected production reactor policymaking over the following decades. Despite the AEC's efforts to soften the economic impact, closing the reactors in groups and seeking local industrial initiatives tended to foster a sometimes adversarial relationship between newly formed grassroots alliances and the headquarters management of the weapons complex. Groups at the two isolated sites regarded the question of reactor policy as intimately wrapped up in community survival and worker job security. The development of political advocates for the affected workers both in Congress and at the local level reflected the growing rejection of the closed-door style of decision-making that the Commission had inherited from the Manhattan Engineer District. In later decades, the agencies that took over the weapons complex from the AEC had to deal with the political groups and alliances stimulated by the AEC's decision to close down 10 of the 14 production reactors.

Inside the AEC, the issues of oversupply and closure forced confrontation with the special nature of management of production reactors. Not only was production of plutonium and tritium controlled by the government as a monopoly, but consumption was all taken by the government, a single-consumer situation that economists call a "monopsony." This unique arrangement within the American economy simply did not fit with the rest of the political-economic structure. The problems of closure reflected the specific difficulties and dilemmas that came from that peculiar situation.

Once a surplus of plutonium was achieved, there was little choice but to take the reactors out of production. President Johnson, in announcing the closures, said that the production reactors could not be a "WPA nuclear project, just to provide employment when our needs have been met."[67] Yet the AEC and local politicians worked diligently to find ways to provide that employment; some of their efforts to maintain employment through new government projects did indeed smack of the New Deal rationale that Johnson formally eschewed.

The dilemma of production reactor availability became clearer by the mid-1960s. Plutonium's half-life is 24,000 years, which meant that, for all practical purposes, once any plutonium was manufactured, it was forever available until fissioned. Shutting down the production capacity was difficult to do on a temporary basis. While some machines could be mothballed and possibly restarted, personnel had to be laid off, and the unique body of skills would erode or disperse. Yet alternative uses for the reactors, such as conversion to electrical power generation or isotope production, proved impractical and uneconomic. To keep the reactors producing plutonium in the face of decreased international tensions, surplus weapons material, and increased hazards was not an appropriate policy decision.

On the other hand, some production reactor capacity had to be maintained for the production of tritium. Although stockpiled amounts of both tritium and plutonium remained classified, it was clear that no matter how much tritium had been accumulated, its half-life of 12 years would reduce the stockpile by half in a little over 10 years if all production ceased. In an era in which the number of tritium-boosted weapons was scheduled to increase, a constant assured production of tritium was required. The reactors scheduled to be kept alive were all capable of tritium production.

In private-sector enterprises, such questions of risk of overexpansion and reaction to the vagaries of market demand were decided at the level of the enterprises. A business could take its losses, alter or shut down an operation, change its product, or possibly, if it could not adjust, go bankrupt as a consequence of a loss of market. In the private sector, the production companies took the risk.

But the production of plutonium and tritium in GOCO facilities represented an anomaly in the American industrial world: government-only monopoly production and government-only monopsony consumption. None of the operating contractors of Hanford or Savannah River risked major capital investments in the enterprises; the contracts provided for cost reimbursement. Demand was not driven by a free or even by a regulated economic market but by the single customer's weapons policy. Policy decisions affecting demand put the government's own capital investment and the jobs of the employees at risk.

Such problems were typical of a "command economy" like that developed by the United States in wartime but were not regarded as typical of the traditional American peacetime economy. It was precisely this aspect of the arrangement that Glenn Lee perceived as a bureaucratic "hammerlock" on the local community of Richland. His complaints sprang from the directly felt local consequences of a national transformation that was felt less painfully elsewhere. As a result of the Cold War and the imperatives of the nuclear standoff, this aspect of the American economy resembled the economy of the Soviet Union, in which decisions were made on a planned basis by a remote government, without reference to market forces, behind closed doors, for reasons that would not be made public. The dedication of a whole community to producing one or two products whose need was secretly set in Washington put the very life of that community at the mercy of the distant bureaucrats, and the sensation was culturally wrenching for Americans.[68]

In the United States, the federal government, after all, was somewhat responsive to political pressures. An AEC decision to lay off 4,000–6,000 workers in a specific locale had immediate political consequences. At Hanford, the response focused through the grassroots community leaders; in both Washington State and South Carolina, governors, state legislators, and members of Congress came to the aid of their distressed constituents. Because the AEC understood the political ramifications, it made the explicit and conscious, but closely held, decision to space out the closures over a period of years, and it made more public the energetic search for alternate projects and products in cooperation with local spokespeople.

But despite the concerns of workers, community leaders, politicians, and contractors, the basic technical problem was not susceptible to an easy political solution. The huge government-owned facilities were practical only for producing certain products; they were appropriate for the purposes for which they had been built, and it was difficult or impossible to convert them to other uses. When the need for one of the products declined, the government faced what ultimately had to be construed as an "on or off" decision when it came to particular reactors. Delaying or stretching out the closings might ameliorate the impact, but ultimately such measures only stretched out the pain and gave the advocates of continued operation more time to organize, to protest, and to build relations with political allies. Placing one or more reactors in temporary shutdown or some form of standby status had only a rhetorical attraction. To keep the reactors truly available required programs of manning, maintenance, training, upgrade, and safety work almost as expensive as continued operation. On the other hand, a true shutdown meant that the capacity would vanish.

From the point of view of the workers and their advocates, alternate nuclear uses for the Hanford site seemed quite appropriate. If other types of nuclear facilities could be built on the site, if a number of corporations could enter the field and hire for a variety of jobs, the Tri-Cities could remain viable.

Eventually, it was inevitable that the United States had all the plutonium it would ever need; conversely, tritium's short half-life meant that production reactor capacity had to be assured with at least one reactor. The remaining four reactors continued to age, inevitably approaching some future date at which their continued operation would no longer be safe. The attempts to come to grips with these production reactor issues in the open forum of national technopolitics in the late 1970s and the 1980s is the subject of the following chapter.

8 • Lobbying for Nuclear Pork

As the surviving members of the production reactor family continued to age through the 1970s and into the 1980s, the AEC and its successor agencies faced the question of where and how to build one or more replacements. Unlike the prompt decisions that led to the design and construction of B, D, and F reactors in 1942–44, and the rapid choice in 1950 of the heavy-water design for the Savannah River reactors, discussion of technological choices for the next generation of reactors dragged on for years. These discussions took place in national forums, including Congress and the national media, and between competing, organized national groups. To focus on this category of reactor technology policy requires a change in perspective that involves looking not only at local scenes and debates, but also at events in Washington. In earlier chapters, technology has been viewed from the bottom up. Here, that viewpoint must be inverted, to look at policy from the top down.

Several factors contributed to the stalemate that characterized production reactor policy over the years 1979–88. Part of the delay derived from the institutional change that replaced the AEC with successor agencies responsible to more congressional committees. Many choices parallel to those that had been reached personally by Brig. Gen. Leslie Groves in the Manhattan Engineer District, or by the AEC in consultation with the General Advisory Committee, the Military Liaison Committee, and the Joint Committee on Atomic Energy, now became of interest to literally dozens of members of Congress. On some of the issues, groupings of citizens, similar to and including those who had fought against reactor closings in the 1960s, but now including organized grassroots proponents and opponents of many sites, entered the dialogues. Issues ranged from fundamental questions of conceptual design, through choice of the appropriate contractor, the weights assigned to site-selection criteria, and the details of reactor safety and environmental impact. Extended debate over such problems engaged dozens of identifiable organizations and thousands of articulate but unaffiliated individuals in what we call "technopolitics." Participatory democracy, in this area, led to technological gridlock, in which decisions were repeatedly postponed.

Between 1972 and 1978 few policy choices confronted those who managed production reactors. As political leaders struggled through the American withdrawal from Vietnam and dealt with the unfolding scandals surrounding the Watergate break-in during the 1972 presidential election, the scaled-down nuclear materials production capacity matched the current and projected needs of the nation. Each year, through the classified Strategic Stockpile Requirements Directive, the president set the amount of tritium required. Productive capacity at Savannah River, and a small backup capacity at Hanford's N reactor, provided the replacement quantities to meet the goals.

On the international level, the strategy of mutual assured destruction, appropriately labeled with the acronym "MAD," had established a nervous stability. Under this doctrine, both the Soviet Union and the United States maintained a sufficient stockpile of weapons to enable each to guarantee that a first strike by the other would be met by a devastating second strike. Meanwhile, the two superpowers followed up on an earlier atmospheric nuclear test ban with a series of treaties in the 1970s that placed limits on the spread or proliferation of nuclear weapons, on antiballistic missile weapons, and finally on the numbers of missiles and launchers (Table 7).

Despite the apparent gains through the treaties of the 1970s, MAD remained the dominant nuclear weapons doctrine for both nations through this period. In a twist of irony, the system of treaties almost ensured that the nuclear arms race would continue and even increase in intensity. Through the Nixon, Ford, and Carter years (1969–80), the stabilized MAD regime and the maintenance of the vast numbers of weapons allowed under the treaty ceilings required steady production of tritium. Multiple independently targeted reentry vehicles (MIRVs) as well as development of the neutron bomb (or "enhanced radiation weapon") also put demands upon the weapons complex for steady tritium production. Critics of SALT I and II noted that the treaties "paradoxically . . . offered irresistible incentives" to continued research and development in the nuclear arms race. Increases in

Table 7. Nuclear arms control treaties, 1963–1979

Treaty	Date	Terms
Limited Test Ban Treaty (LTBT)	10 Oct. 1963	Signatories agreed not to test nuclear weapons in the atmosphere, in outer space, or under water
Non-Proliferation Treaty (NPT)	5 Mar. 1970	Signatories agreed not to aid non-nuclear powers in developing nuclear weapons
Antiballistic Missiles (ABM)	3 Oct. 1972	U.S. and USSR agreed to limit anti-ballistic missile sites to two each
Strategic Arms Limitation Treaty I (SALT I)	3 Oct. 1972	U.S. and USSR agreed to limit number of missile launchers
Threshold Test Ban Treaty (TTBT)	3 July 1974	U.S. and USSR agreed to limit underground tests to 150-kiloton devices
Strategic Arms Limitation Treaty II (SALT II)	18 June 1979	U.S. and USSR agreed to ceilings on strategic weapons delivery systems

Source: Julie Dahlitz, *Nuclear Arms Control* (London: Allen and Unwin, 1983), 24–31.

planned weapons systems in the late 1970s—including cruise missiles, a planned mobile-launcher system (MX), and missiles for the Trident submarines—all demanded a large plutonium supply, putting an end to the plutonium glut and even requiring increased production of that material.[1]

Even had there been no increase in U.S.-USSR tensions and no new weapons systems, the arms race required continued tritium production to keep up with the erosion of supply due to the isotope's short half-life. Hence, reliable production reactors were required for both strategic materials. The remaining production reactors, built in the 1950s, were aging and less reliable. However, the nuclear establishment of the 1970s and 1980s operated by quite different rules than the institutions which had faced similar choices of site, technology, and contractor in the 1940s and the 1950s.

The Widened Political Environment

Congress reconfigured the nation's nuclear administration in the early 1970s. A variety of energy issues through the early 1970s, including electrical power blackouts and "brownouts," and concerns over dependence on imported oil, led to support for a coordinating energy agency. Furthermore, criticisms mounted that the agency that promoted the rapidly expanding nuclear power industry should not also regulate that industry through licensing and inspection. Congress ended the life of the Atomic Energy Commission in 1974, transferring the production facilities to the new Energy Research and Development Administration (ERDA), later replaced by the Department of Energy (DOE) in 1977, and regulatory and licensing powers over commercial reactors to the Nuclear Regulatory Commission (NRC). Congress decided that government-owned reactors should not be under NRC jurisdiction, so ERDA (and then DOE) retained management and control of the production, test, and experimental reactors formerly owned by the AEC. Congressional oversight by the Joint Committee on Atomic Energy was abandoned, and aspects of jurisdiction over the weapons complex were delegated to at least seven separate committees: the Senate Committee on Governmental Affairs, the Senate Committee on Energy and Natural Resources, the House Committee on Interior and Insular Affairs, the House Committee on Energy and Commerce, the Senate Finance Committee, the House Ways and Means Committee, the House and Senate Armed Services Committees, and from time to time, specialized subcommittees in both houses.

Since so many House and Senate committees became involved in oversight, the administrator of ERDA, and later the secretaries of energy, had to be equally concerned with several powerful committee chairs, rather than just one, as in the days of the AEC-JCAE relationship. When a new production reactor was planned, political advocates of various sites had many more congressional venues for attempting to influence the choice of sites. Even after a site was chosen, advocates of other potential sites could use their representation in Congress to delay action and request further study. The number of senators and representatives directly affected by reactor siting had become quite large.

Early in reactor development, the AEC had sought a more isolated site for experimental reactors than the small county forest preserves just west of Chicago that were used by the Argonne National Laboratory. Thus, in 1949 a 400,000-acre tract near Idaho Falls, Idaho, had been established as the Nuclear Reactor Testing Station; it was redesignated in 1975 as the Idaho National

Engineering Laboratory (INEL). This site, which at 890 square miles was even larger than the 590-square-mile Hanford reservation, became the location for experimental breeder reactors, materials and process testing reactors, and prototype submarine and ship reactors under the Navy's jurisdiction. From its first days, the Idaho facility housed work done by a variety of nuclear contractors and laboratories. Construction of a total of 52 reactors at the Idaho site enriched the experience in nuclear engineering there, since these reactors represented about one-third of all federally owned test and experimental reactors ever built in the United States. Over the decades, INEL, like Hanford and Savannah River, had developed a base of trained reactor personnel; and the state of Idaho, like Washington and South Carolina, had developed a pro-nuclear community of local and regional political representatives and spokespeople. The 52 reactors included 11 built by Argonne, 11 by General Electric, 11 by Aerojet General, 11 by Phillips Petroleum, and smaller numbers by Westinghouse, General Atomics, and Combustion Engineering. By the 1980s, INEL's network of congressional friends included not only representatives from Idaho but a few from other states in which the contractors were headquartered.[2]

Thus, when a new production reactor was considered, senators from Washington and South Carolina were joined by colleagues from Idaho in contending for possible siting of a reactor. Six members of the House of Representatives from South Carolina, seven from Washington State, and two from Idaho, as well as governors and state legislators, added their voices to those of the senators. All focused their pressure through a number of congressional committees, each well-informed on nuclear matters. Because of the seniority system, several of the six senators from these states were men of considerable power in the 1980s (see Table 8).

In the 1980 elections, not only did the Republican candidate, Ronald Reagan, win the presidency, but Republicans also gained a majority in the Senate. The effect of this change was profound; Republican senators assumed committee chairmanships and other positions of power. Strom Thurmond of South Carolina became president pro tem of the Senate; James McClure of Idaho became chair of the Senate Committee on Energy and Natural Resources. These men in particular used their newly won positions to influence the development of a new production reactor in the 1980s. Each preferred the combination of site selection criteria and technology choice most likely to bring the reactor and its attendant employment to their home districts.

In addition to the powerful senators, outspoken members of the House of Representatives also worked to advocate the interests of the states that

Table 8. Relative seniority of U.S. senators from reactor-site states

State/senator (party)	Senate tenure
South Carolina	
Strom Thurmond (R[a])	1957–
Ernest Hollings (D)	1967–
Washington	
Warren Magnuson (D)	1945–1981
Henry Jackson (D)	1953–1983
Slade Gorton (R)	1981–1987, 1989–
Brock Adams (D)	1987–1993
Daniel Evans (R)	1983–1989
Patty Murray (D)	1993–
Idaho	
Frank Church (D)	1957–1981
James McClure (R)	1973–1991
Steven Symms (R)	1981–1993
Larry Craig (R)	1991–
Dick Kempthorne (R)	1993–

Note: [a]Thurmond changed party affiliation from Democrat to Republican in 1964.

Source: U.S. Senate Historical office.

had potential sites for production reactors. Butler Derrick of South Carolina, an articulate spokesman for the district that included the Savannah River Site, served from 1975 through 1989. Mike McCormack, representing the district in Washington State that included Hanford, served from 1970 through 1981. Prior to his election, he had worked for 20 years as a research scientist at Hanford. Sid Morrison, who followed McCormack, was an active advocate of the Tri-Cities area and its nuclear interests. Senators and representatives from Georgia joined their colleagues from South Carolina in expressing concern over the impact on employment in the region.

Calvert Cliffs, Ralph Nader, and Three Mile Island

Beyond the formal political structure, informal groups of citizens, organizing around a variety of issues, also played a role in advocating or opposing the construction of a particular new production reactor at a particular place. Some of the organizations were themselves a product of the crosscurrents in American society which came to the surface in the 1960s and early 1970s.

Court decisions, new and articulate leaders of consumerism and environmentalism, and events in the world of commercial reactor cousins all contributed to the new constellation of forces that affected the process of selecting a site and technology for the new production reactor.

A major step in the growing public participation in reactor technology decision-making derived from a court case involving the siting of a Baltimore Gas and Electric nuclear reactor plant in Maryland. A federal district court ruled in the 1971 Calvert Cliffs decision that henceforth, National Environmental Policy Act requirements for full environmental impact statements (EISs) and public hearings would apply to the nonradiological impacts of new power reactors, while an abbreviated study still sufficed on the strictly radiological impacts. Then, in 1972–73, rule-making hearings regarding emergency core cooling systems brought to national media and public attention the simmering issue of reactor safety.[3]

With the sale of reactors to more and more utilities, and the consequent playing out of nuclear reactor siting determinations before an ever-widening audience, the issue of the measurement of reactor risk moved more squarely into the public eye. Although nuclear engineers had discussed and studied the probabilistic risk assessment (PRA) approach in the late 1960s, the general public was first introduced to the concept with the publication in 1974 of a controversial report. The report, authored by a group headed by MIT professor Norman C. Rasmussen, was published first by the AEC (WASH-1400) and then in final form by the NRC (NUREG-75-014). The report stimulated widespread study and usage of probabilistic methods, as well as adverse publicity from critics who saw it as a biased and unabashed defense of power reactors. Such criticisms were fueled by a statement in the preface of the first edition, that the risk of death from a nuclear reactor accident was less than that of being struck by a meteor, which had a degree of mathematical truth. The timing of the report also seemed to critics to substantiate the charge that it was not scientific, but political. The report had been commissioned to coincide with congressional discussion of reenactment of the Price-Anderson bill, which limited the liability of utility companies for claims arising from commercial reactor disasters. Partly because of the glib comment about meteors, critics viewed the Rasmussen report and its probabilistic approach as an effort to allay popular fears and secure renewal of the liability limitation. Its timing, its language, and its sponsorship struck opponents as proof of its political bias and led them to discount its claims to provide technically objective methods of safety evaluation. In actuality, the models employed by Rasmussen were a step forward in establishing some means of estimating reactor risk. The NRC recognized that the PRA method allowed for a numerical measure of overall reactor risk and began to demand PRAs of reactor sites and reactor technologies over the next years.[4]

The reception of the Rasmussen report showed the extent to which the judgment of technical experts came to be treated in the public eye as a political or polemical issue by the late 1970s. That change from acceptance of expert opinion as objective to treating it as subjective judgment, motivated by ideological commitment, was exemplified and articulated by the founder of popular consumerism, Ralph Nader.

"The tide is turning," Nader and John Abbotts wrote in their 1979 book *The Menace of Atomic Energy*. They pointed to a rising public movement that based its criticism of nuclear energy on several profound issues. Nader and Abbotts traced the origins of the antinuclear movement to the government's own efforts to stimulate a nuclear electric-power-generating industry, which, they claimed, had been "insulated" from public view intentionally. In the post-Watergate, post-Vietnam age, in which public suspicion of politicians and experts had mounted to new levels and in which demands for participatory government flourished, the AEC's initial method of promoting and developing nuclear power out of public view had itself created the opposition.[5]

Nader and Abbotts pointed to a host of more specific concerns related to the nuclear power industry about which the public had insufficient information: taxpayer subsidies to the industry, guarantees against nuclear theft and sabotage, the Price-Anderson bill's limited liability protection for industry, faulty emergency evacuation plans, worker exposure to radioactivity, and decommissioning costs. They decried "sweetheart standards" in which the very agency advocating nuclear power had set the rules for the industry. Specific issues such as thermal pollution of lakes and streams by cooling water were troubling, and these authors appealed to concern for fisheries and for a protected environment in their arguments against power reactors. The electric atom, they claimed, developed from the wartime secrecy surrounding nuclear weapons, and it never met a "market test, an open information test, an electoral test, or in fact, much of a Congressional test."[6]

Proponents of nuclear power met many of Nader and Abbotts' criticisms with cogent counterarguments. Much of the secrecy was necessary by definition, such as that regarding protection against theft, sabotage, and emulation of technology by potential nuclear states. By its nature, public information was incompatible with public safety in such areas. Although placing limits on liability passed the risk of catastrophe to the potential victims, rather than sharing it as an insurance cost to all consumers or taxpayers, that pattern resembled catastrophic risk exposure of the same scale, as in cases of natural disasters like earthquakes, floods, or hurricanes. Thermal pollution of lakes and rivers could be mitigated. Health hazards from other energy systems, such as fossil-fuel electric plants, were severe during regular operation. The Atomic Energy Acts of 1946 and 1954, the National Environmental Policy Act of 1970, the Energy Reorganization Act of 1974, and the act creating the Department of Energy in 1977 had all been thoroughly aired and debated. Yet Nader and Abbotts had captured the sense of frustration that had come to focus in the anti-nuclear power movement by the late 1970s; they understood very well the sources of that frustration and hostility and expressed the points clearly.

The frustration, hostility, and fears regarding reactor risk that Nader and Abbotts had identified suddenly erupted on a much larger scale in response to a nuclear power reactor accident in Pennsylvania in 1979. In April, Unit 2 of the Three Mile Island reactor station experienced a loss-of-coolant accident. A stuck-open pressure relief valve and operator error compounded by confusing control-room signals worsened the accident. A partial meltdown of fuel resulted; fortunately, the containment structure prevented release of most radioactivity to the environment. Millions watched the unfolding story in continuing television coverage. This highly publicized incident, coming against the background of heightened public concern over commercial power reactors, gave substance to the arguments of those who agreed with Ralph Nader. A somewhat less well known effect of the incident was to stimulate further interest among nuclear engineers in PRA methods, which could evaluate the risk from precisely such unlikely combinations of individually unlikely occurrences as the failure of minor parts, poor communications, and operator error.

The Three Mile Island incident also helped to stimulate membership in organizations that attempted to bring together those opposed to nuclear power with those advocating disarmament. The Mobilization for Survival, or "The Mobe," formed in 1978 and composed of some 40 national and regional groups, immediately began to grow and to mount a series of demonstrations against parts of the weapons complex and against nuclear power reactors. This phenomenon of "convergence" between peace groups and antinuclear groups grew out of the desire of both movements to develop broader constituencies.[7]

These developments in the politics of commercial reactor cousins slowly began to affect the shrinking world of the remaining production reactors. There were several reasons that production reactors could continue to operate in relative isolation from public scrutiny. First of all, there were only four production reactors in operation in the late 1970s (P, K, and C at Savannah River and N at Hanford) in contrast to the more than 70 commercial power reactors. Although the function of production reactors was not classified, their specific output and their day-to-day functioning were. Their routine operation required no new decisions. Furthermore, the four operating reactors were located within huge federal land reservations devoted to the nuclear weapons complex. Access to the Hanford and Savannah River facilities was strictly controlled and limited to employees and others on official business. Photographs of any of the production reactors were available only through official sources. By contrast, commercial reactors normally had much smaller "exclusion zones" than the production reactors, and many were clearly visible from well-traveled highways. The production reactors had been built decades before; all the commercial reactors were of more recent vintage, and decisions about their licensing and siting were commonly in the news.

However, when one or more production reactors closed, the public attention was drawn, as it had been in the earlier round of closings, to the issue of employment. The same organizations that had mobilized to protect jobs in the 1960s continued to operate. Their friends in Congress continued to represent them, now with the opportunity to speak through many more influential and strategically situated committees that ruled on DOE's budget and policy. And when decisions had to be made regarding a new production reactor, public focus on the issue of siting evoked a simultaneous scramble for the employment and related business expenditures from advocates and a set of protests from the groups and interests who had opposed the expansion of commercial power reactors. Opposition was voiced through loosely joined networks of organizations such as The Mobe, as well as through local organizations such as those among the "downwinders" in Washington State. The growing public distrust of the objectivity of nuclear experts added to the arguments of those who sought public review of production reactor decisions.

Thus, by the end of the 1970s, the straightforward issue of siting a new production reactor (NPR)[8] could not be resolved as simply and quietly as when General Groves selected Hanford. The technically complex issues of conceptual design choice that had spawned a heated debate in the top-secret P-9 Committee between Eugene Wigner on the one side and Crawford Greenewalt on the other in 1943 were paralleled in the debates of the 1980s, although such issues were less readily understood by journalists and the general public. But bringing the conceptual design debate out from behind closed doors into the open allowed the corporate advocates of differing systems to seek political support. The *New York Times*, the *Washington Post*, *Newsweek*, and television news reporters described reactor designs; members of Congress found themselves reviewing the merits of the various reactor technologies. Usually such an interest allowed advocates of one site to focus on the *demerits* and disadvantages of technologies that were linked to the other competing sites, in hopes that their own site would benefit. In such a fashion, local economic self-interest, corporate lobbying, and pork-barrel politics contended in site against site, and technology against technology, as well as against a medley of voices motivated by concerns for the environment, distress at radiation risk to downwind residents, nuclear disarmament, democratization of decision making, and distrust of experts.

Yet it was not simply the opposition of antinuclear advocates that stalled progress towards the construction of the next generation of production reactors. Rather, over the period 1980–88, congressional representatives from the states of Washington, Idaho, and South Carolina vied for increased funding of atomic energy facilities. Ironically, it was this contest between the various interests *advocating* a new production reactor that brought the planning effort to a standstill through this period.

The battles over the NPR through the 1980s centered on a series of published reports by selected groups of experts. Although the secretaries of energy hoped to get from prestigious experts objective and technical evaluations, the groups' reports themselves became politicized. Like the study by Rasmussen, these attempts at objective evaluation came to be treated by the various advocates as sources of arguments for or against a particular technology and site. If a group of experts advocated expenditure or planning without specifying site or technology, proponents of different reactor designs and locations could all agree upon the report as an argument for action. However, if members of a technical panel favored one choice over others, opponents criticized them as biased. If the experts pointed to the relative demerits of various plans, those demerits provided reasons for not proceeding. The cycle of expert advice and countersuggestion repeated itself several times over the period.

Reports and Recommendations

In 1980, several policy groups concluded that production reactors needed to be upgraded or replaced, laying an objective base of need for a new production reactor without showing a preference for a particular site or a particular technology. The Department of Energy submitted to Congress in 1980 a study that evaluated the need for restoration and replacement of parts of the deteriorating nuclear complex over the following 5 years. In the same year the National Security Council also concluded that production facilities planning must address increased materials requirements, aging facilities, and new production capability.[9] On 11 April 1980, Secretary of Defense Harold Brown expressed his concern over DOE's possible inability to meet future strategic materials needs in a letter addressed to the secretary of energy.[10]

On 15 July 1980, the Department of Energy/Department of Defense Long-Range Resources Planning Group concluded more specifically that one and possibly two new production reactors were needed by the year 2000; the group recommended a new production reactor on line by the 1990s.[11] After urging from Republican critics, and after leaks of internal memoranda about Defense Department concerns, President Jimmy Carter approved expansion of plutonium production on 25 September 1980, but without specifying where or how to increase the production. To an extent, his concern for increased plutonium production stemmed from heightened tensions over the 1979 Soviet invasion of Afghanistan; his announcement may have had the effect of offering a diplomatic signal of American resolve.[12]

In 1981, a DOE evaluation determined that existing reactors were unlikely to meet defense needs by the 1990s and that planning was needed for a replacement production reactor.[13] Anticipating that the Idaho National Engineering Laboratory could be a site for the next production reactor, DOE established a replacement production reactor project office there.[14]

In response to all these generic recommendations, in December 1981 Congress enacted Public Law 97-90, which appropriated $10 million to the DOE's Office of Defense Programs under project 82-D-200 for continued study of the need for an NPR at an unspecified location.[15] Congress agreed that a reactor was needed and left to the experts the issue of study as to type and location. The work was to be limited to architectural and engineering

(A&E) studies only. Under section 201 of the act, none of the designated monies could be reprogrammed to fund any other sort of work without concurrence from "appropriate committees" of Congress or, failing explicit concurrence, the expiration of a 30-day period of notification to those committees. This provision limited the secretary of energy to technical studies of a new production reactor at an unspecified site, unless he sought approval to redirect the funds, providing a very tight system of congressional oversight. In this fashion, the various congressional interests representing the three sites could agree that work should progress and postpone temporarily any battle over exactly who should get the prize.

Within a year of his inauguration, President Ronald Reagan, following the concerns about strategic materials shortfalls enunciated by his predecessor, publicly announced his support for expansion of both tritium and plutonium production. In announcing the president's commitment, Deputy Assistant Secretary of Energy F. Charles Gilbert remarked on the alternatives for increasing production that had been discussed, including the restart of closed reactors and retooling of N reactor. He denied rumors that a new production reactor had been promised to the Idaho site, rumors spurred by the creation of the replacement office there and by the fact that Idaho was represented by a Republican senator.[16]

Over the period from 7 July 1982 through 2 November 1982, DOE sought advice through the specially appointed Concept and Site Selection Advisory Panel (CSSAP) convened under the chairmanship of former Atomic Energy Commissioner T. Keith Glennan.[17] The Glennan panel studied seven possible technologies:

HTGR	high-temperature gas-cooled reactor
LMFBR	liquid-metal fast breeder reactor
LTHWR	low-temperature heavy-water reactor
LWR	light-water reactor
RNR	replacement N reactor (graphite-moderated, water-cooled)
WNP 4	Washington Nuclear Power Project no. 4 (conversion of partially completed LWR power reactor)
ZEPHR	zero-electric-power heavy-water reactor

Technical advocates of some of the proposed systems were better organized than others. Du Pont continued to represent the heavy-water technology but with little enthusiasm for expansion of its reactor business. Light-water-reactor technology attracted Westinghouse, which had developed considerable experience with pressurized water reactors over the 1970s for commercial power applications. General Atomics advocated the high-temperature gas-cooled reactor. That corporation, a former division of General Dynamics acquired by Gulf Oil in 1967, had participated in the design of the Peach Bottom no. 1 40-MW(e) gas-cooled reactor, which went critical in 1966, and had designed and built the larger Fort St. Vrain 330-MW(e) gas-cooled reactor, which went critical in 1974. General Atomics actively promoted its reactor designs, working through several consortiums of industrial groups. Moderated by graphite, cooled by high-temperature helium, and fueled by uranium in ceramic-metallic pellets, the HTGR design was regarded as "inherently safe." European experimentation with HTGRs through the 1970s had advanced the state of the art there; General Atomics could claim, with some justification, that the United States should fund gas-cooled technology if only to stay competitive in the developing nuclear technologies.[18]

The proposal to convert a partially completed WNP reactor to plutonium production had another group of advocates. The Washington Public Power Supply System consortium, now composed of 88 electrical cooperatives and municipalities, worked through political representatives in Washington to present the conversion of such a reactor as a financial boon to the region and to the system. Yet Rep. James Weaver of Oregon and Sen. Gary Hart of Colorado, among others, argued against the conversion, on the grounds that such a move would destroy the historic distinction between civilian and military nuclear programs.[19] The Glennan panel was well aware of this line of objection to the WNP conversion concept.

Although advocates of one technology or another were often associated with locales or corporations that had experience with the particular designs, unaffiliated experts often found themselves drawn into the technopolitical dispute. There were many articulate advocates of the HTGR outside General Atomics, for example. In a work critical of the AEC's decision to proceed with studies of the liquid-metal fast breeder reactor, Thomas Cochran, writing for the Ford Foundation–funded Resources for the Future, listed a number of superior features of the HTGR that made it a safe reactor and a potential electric power source.[20] General Atomic, the leading gas-cooled reactor advocate, was a relative latecomer to the commercial power reactor business and had fallen behind its major competitors, General Electric and Westinghouse, who had entered the business during the period of greatest power reactor construction, in the early 1960s, with their light-water models. Both General Atomics and the HTGR design proponents saw in the new production reactor effort a chance for the HTGR technology to be developed, at a time when support for this technology had waned in the private power reactor competition.[21]

In addition to evaluating possible technologies, the Glennan panel also reviewed four possible sites—Savannah River, INEL, Hanford, and the Nevada Test Site, the nuclear weapons testing facility north of Las Vegas, which was pockmarked by hundreds of underground blasts. Glennan immediately rejected Nevada as a possible reactor site because of earthquake hazards. In a process similar to that used by Groves in 1942 and by Du Pont in 1950, Glennan used 11 criteria to evaluate sites, including water supply, transportation facilities, available labor, and support facilities. New criteria included waste disposal capabilities, environmental impact, public acceptance, and the "duality" of site. The last factor referred to the desirability of having two sites operating at the same time; since it was assumed then that older reactors at Savannah River would continue to operate into the 1990s, the other sites had a slight edge for location of the new reactor in this regard.[22]

When he submitted the report, Keith Glennan reminded Secretary of Energy Donald Hodel that the subject of a site and a technology for a new production reactor had already been studied several times. Glennan explicitly drew attention to "the proprietary interests of many of the companies and Congressional people whose constituencies may be involved in each of the concepts or sites dealt with." The Glennan panel had attempted to deal with these interests, he said, in an even-handed and thoughtful way. "Further studies," he warned, "will do little to increase the validity of the recommendations our Panel has made."[23] Although correct and prophetic, his warning did not prevent further studies.

The Glennan Report, or more formally, the "Report of the New Production Reactor Concept and Site Selection Advisory Panel (CSSAP)," supported the general idea of an NPR by the 1990s as necessary to "assure an adequate supply of strategic materials." All the advocates of various technologies liked that part of the report. However, the panel then spelled out its choices. CSSAP favored as a first choice a heavy-water reactor along the lines of those at the Savannah River plant, with no power generation—that is, the ZEPHR; as a second choice, the panel favored either a LTHWR or a replacement N reactor. The Glennan Report specifically discounted the other technologies, including a light-water-cooled and -moderated version, the high-temperature gas-cooled graphite-moderated (HTGR) model, and a liquid-metal-cooled fast breeder. The report noted that exploration of those technologies that had been considered for power production might raise worldwide concerns that the United States was planning to convert commercial reactor designs to weapons material production, reflecting the line of thinking of Representative Weaver and Senator Hart. Furthermore, Glennan personally doubted whether the attempt to develop an N type dual-purpose reactor with steam sale to offset cost was a good idea; its high initial cost could further "lower the probability of NPR program approval." On the issue of location, Savannah River was preferred, Hanford ranked second, and INEL was disapproved.[24]

Glennan's attempt to present his committee's findings as purely objective immediately became undermined, as all the advocates of the choices his panel ranked as poor came to the defense of their own proposals. In addition to the corporate interests behind light-water technology, heavy-water reactors, and the HTGR, political representatives defended Hanford as a location for the new production reactor or for operation of WNP, Idaho as a location for any of the designs, and Savannah River as ideal for the heavy-water designs. Other arguments for Idaho included the site's general experience with more than 50 experimental reactors, the local tradition of work on gas-cooled reactor technology that went back to experiments with mobile reactors in the 1960s, and the group of friendly associated firms and institutions such as EG&G and Argonne National Laboratory.

Predictably, the only congressional delegation that read the Glennan Report with much favor was South Carolina's, and the only corporate support came from the group involved in heavy-water work, Du Pont. Senator Thurmond offered his thoughts to Secretary Hodel. In particular, Thurmond wanted further information on the proposed cooling system, potential effects on the Savannah River, and details regarding radioactive waste management. Thurmond urged Hodel to move along with implementation, but suggested that environmental details needed attention.[25] Hodel reassured Thurmond that, whatever site was chosen, construction would be in full compliance with the National Environmental Policy Act, and he promised to provide the needed information as studies proceeded.[26] Following up on the Glennan Report, in December 1982 Du Pont released a management plan for the first phase of a potential NPR project at Savannah River that included environmental and safety analyses and conceptual design studies.[27]

While experts and politicians worked on the issue of a replacement or new production reactor, the tritium shortfall remained. Thus, while a long-range solution was being sought and argued through, Congress approved a short-term solution, a restart of one of the previously closed Savannah River reactors. Debates over the restart issue proceeded simultaneously with the Glennan review and its response.

The L Restart Controversy

In the summer of 1980, as a temporary way of meeting plutonium and tritium requirements, Congress authorized funds for the restart of production reactors previously shut down by President Johnson. This funding allowed investigation into the condition of L reactor at Savannah River and the planning for its reopening over the objections of a variety of environmental and disarmament advocates.[28] DOE's approach was to launch an internal study of all the environmental issues—including thermal discharge, radiological dose, and environmental surveillance planning—without following the EIS procedure of holding open hearings. Then, using the appropriated funds, the reactor would be repaired and put in production. In order to allay public concerns over the lack of the full EIS procedure, Senator Thurmond held a round of local hearings as part of the Armed Service Committee's responsibilities. These February 1983 hearings over the L restart provided insight for Congress of how production reactors had become a focus of so many varied political interests and showed the popular side of the struggles involved.

In these hearings Sen. Mack Mattingly, whose constituency in Georgia lay along the Savannah River, expressed concern for the environmental and public health impact of the restart, both in terms of the thermal effect on the river and in terms of possible radioactive releases, and wanted a full environmental impact study prepared. Nevertheless, Mattingly supported the need for the reactor for defense purposes and suggested that anyone who wanted to turn the hearings "into a debate over disarmament" should leave. South Carolina representative Butler Derrick, while supportive of the restart of the reactor, believed that DOE had erred by not encouraging public participation in the environmental assessment process. After hearing from Troy Wade, deputy assistant secretary for Defense Programs, and from other DOE personnel, a representative of the state of South Carolina added the voice of the state governor to the call for a thorough public hearing on the environmental issues.[29]

Among concerned and self-proclaimed experts, the committee heard from representatives of the South Carolina Wildlife Federation and the Sierra

Club, who warned of negative impacts; from a former Du Pont employee who insisted that an EIS would be a waste of funds; and from a representative of the National Academy of Sciences, who gave data suggesting that L reactor had never created any problems for the natural organisms in the river. A pediatrician warned against the health effects on children. A representative of Physicians for Social Responsibility took a strong stand against further plutonium production and denied the concept of a maximum low safe level of radiation, or a "threshold" of exposure.[30]

The ten Aiken County representatives in the South Carolina state legislature were unanimously in favor of the restart. The League of Women Voters sent three representatives, all of whom argued for a thorough environmental impact study. The mayors of Aiken, Augusta, North Augusta, Williston, and Allendale reported that their communities were clear in support of the restart. Thomas Cochran, representing the Natural Resources Defense Council, vividly described the problems of restarting L reactor, closed since 1968. Pigeons had roosted in the rusting equipment, while weeds grew on the grounds. He alluded to routine radioactive releases and a variety of other problems that suggested L could never meet the standards set for commercial reactors.[31]

Representatives of chambers of commerce, wetlands groups and water authorities, the Georgia Conservancy, labor unions, and the Coastal Citizens for Clean Energy all expressed varying degrees of concern over employment and environmental issues. One representative, Michael Gooding from the Grass Roots Organizing Workshop, reminded the senators, over objections from Thurmond, that the government that wanted to restart L reactor was "the same Government that . . . was willing to napalm children, women and old men in Vietnam, and is now willing to support the massacre of people in Central America." The government, he said, was in the employ of the ruling class and would "do whatever it pleased." A less impassioned representative of the World Affairs Council of Georgia State University also criticized the arms race itself; several other unaffiliated speakers criticized the nuclear industry's willingness to accept public risk. Senator Thurmond and other advocates of work on L reactor patiently sat through the range of hostile critiques.[32]

The heated struggle at the L restart hearings revealed the sorts of arguments production reactors could evoke by the early 1980s. If any NPR were to be finally proposed for one of the sites, such a machine, like L reactor, could serve as a focal point for all of the arguments over risk, peace, safety, endangered species, threshold of radiation exposure, and the role of the United States in world affairs. Not only would the sites vie with each other for the benefits, but dozens of organized groups would raise objections.

As to L reactor, for a short period the aged reactor rejoined her sisters in meeting the demands of the arms race. The Department of Energy issued a final environmental impact statement in 1984, then remodeled and reconditioned the reactor.[33] L restarted in 1985; it operated for less than 3 years, closing in 1988.

Roadblocks to the Glennan Recommendations

While Secretary Hodel was still evaluating the Glennan Report, Sen. James A. McClure, the Republican senior senator from Idaho and chairman of the crucial Energy and Natural Resources Committee, wrote a detailed, technically well-argued, and lengthy letter to Hodel giving his opinion on the report. Senator McClure questioned Glennan's assessment of INEL as an inferior location for a new production reactor. In particular, McClure argued that four of the selection criteria should have been given more weight: the reactor "need date," the duality of sites issue, issues of new technology, and lowest life cycle cost. By setting the need date artificially close to the present, he argued, Glennan ended up giving preference for a tried and true technology that was 30 years old—the heavy-water models of the 1950s. If there was less of a rush, he argued, the department could take the time to develop a more innovative technology.[34] Of course, once these factors were weighted as McClure preferred, then INEL seemed like the first choice, since it offered "duality" with either Hanford or SRS, since it was not committed to either the light-water/graphite models of Hanford or the heavy-water models of Savannah River, and since it had been the site for experimental work on reactor types.

Senator McClure's need for an "objective" report that would justify INEL as a location rather than Savannah River was met in 1982. The President's Office of Science and Technology Programs released a study that confirmed the need for an NPR and suggested that future studies focus on three technologies: the high-temperature gas-cooled reactor (HTGR), the pressurized-water reactor (PWR), and a replacement N reactor (RNR) at Hanford or INEL. To set up a method for implementing the studies, on 22 July 1982, Los Alamos National Laboratory published a study entitled *Proposed Activities and Funding Requirements for the NPR Program Requirements Office*, which discussed tasks, funding, and the provision of technical support to the project.[35]

In January 1983, DOE established a project charter for an NPR.[36] The department set up within the office of Defense Programs a "desk," DP-13 (later redesignated DP-132), which became the focal point for future planning of an NPR to meet the clearly established need for an assured source of tritium. Yet the tension between the advocates of Hanford, INEL, and South Carolina remained; and the department was unable to secure funding without a break in the congressional deadlock. The small staff at DP-13 was literally swamped by the generation of paperwork and evaluations over the next few years. The office spent a sizeable proportion of its budget, first with a unit of the consulting firm EG&G and then with a Maryland-based branch of the Argonne National Laboratory, to provide office support services, which concentrated on gathering the documentation flooding into the office. Internal staff and outside contractors evaluated locations and technologies, absorbing a budget of $10 to $20 million per annum in these activities. The files collected by the two office support contractors exceeded 75 linear feet by 1988.[37]

As the information came in, however, Secretary Hodel moved rather prematurely to try to force the issue. On 9 August 1983, by internal memorandum, Hodel directed staff to develop a final site and concept recommendation to deliver to President Reagan within the following 18 months. Hodel's personal ties to the Northwest—he had been a Bonneville Power administrator and was a native of Oregon—may have disposed him favorably to the arguments in favor of the Idaho site. He evaluated the Glennan Report, accepting its recommendation of a heavy-water reactor as a tested technology but indicating his "current preference" was to locate the reactor at INEL. His reasoning was that this site choice guaranteed duality of location, following this element of McClure's complaint. He directed that an EIS be developed encompassing an assessment of the environmental impact of the reactor, a risk analysis, a study of socioeconomic impacts, a survey of endangered species, a study of transportation, a hazardous waste management plan, and an archaeological survey. He anticipated that the tritium requirement in the plan should indicate that a completely new "standard" reactor was required by 1995.[38]

Hodel expected to present to the president within 18 months a recommendation for a decision, based not only on the proposed environmental study, but also upon further study of developmental issues related to the technologies. He asked Defense Programs to use currently appropriated funds to work as quickly as possible in conducting the studies and asked that the "management-by-objective tasks" for the new production reactor be altered to include new "milestones" reflecting his decision. Hodel's intention, couched in governmental management language of the early 1980s, was clear to those in the agency: he expected them to set specific goals in preparing the studies and to move promptly towards those goals.[39]

Technically, however, to conduct the studies he requested involved using funding for environmental work that had been set aside by Congress for architectural and engineering studies. This funding had specifically not been appropriated for environmental work, especially pertaining to a particular site. Such a reallocation of funding presented a stumbling block, for a shift of funding to environmental work opened the issue to review by a variety of members of Congress.

On 16 August 1983, Hodel officially notified the House of Representatives Armed Services Committee chair, Melvin Price, of his intent to prepare an EIS for the NPR, but in his request he did not specify which site he preferred.[40] He requested immediate approval of the reallocation of funding to proceed with the EIS, even in advance of the 30 days allowed to approve or disapprove any reallocation. This request had several immediate political effects.

Hodel's actions angered and aroused the political delegations from South Carolina and Washington, who felt that the preference for Idaho announced internally by Secretary Hodel was premature and represented an ill-informed decision, especially since Glennan had specifically discounted that site. Secondly, a host of grassroots organizations supportive of locating the reactor in Idaho launched concerted campaigns. And thirdly, other organizations actively spoke out against siting the NPR at INEL. The advocates focused on skills, employment, and economic benefits; opponents to the planned siting of the reactor in Idaho echoed the arguments against commercial reactors, focusing on environmental and risk questions; some opponents raised the issue of the morality of the nuclear arms race. Some of the participants directly echoed the arguments being made at about the same time at the L restart hearings.

Secretary Hodel's known preference for Idaho as a site for the new reactor, while pleasing to Senator McClure, immediately aroused the ire of other highly placed senators. Less than two weeks later, on 25 August 1983, Senator Thurmond told Hodel that he wanted him to reconsider Savannah River as an NPR site and to delay any final decision on location. Thurmond pointed out that other aspects of the nuclear weapons complex did not have the duality-of-site location that Hodel was using to justify a location other than Savannah River. "I would hope," Thurmond said, "that once again you may be persuaded to accept the findings of the experts, and conclude that [Savannah River] is the most desirable location." Thurmond relayed letters from his constituents, as well as correspondence from the South Carolina state legislature.⁴¹

On 6 September 1983, Sen. John Tower of Texas, chairman of the Senate Armed Services Committee, told Hodel he approved the preparation of an EIS for a new production reactor with the contingency that *all three* sites receive equal consideration and that the environmental consequences of at least two technologies be examined prior to final selection of location or type.⁴² A week later, however, Melvin Price, of the House Armed Services Committee, indicated that he did not think an EIS was called for at that time, since the study itself would cost several million dollars. He asked Hodel to testify for the next budget on all the "factors, contingencies and alternatives under consideration for a facility which cost billions of dollars."⁴³ In November 1983, Hodel responded to Tower's concerns by announcing DOE plans to conduct a series of studies on the need for and cost of an NPR.⁴⁴

The reception and handling of the Glennan report revealed the nature of the political deadlock. Idaho's McClure effectively blocked the Glennan preference for siting in either Washington State or South Carolina. Then representatives of both Washington and South Carolina stalled the implementation of the Hodel concession to McClure. The dispute demonstrated that congressional representatives could at least prevent each other from getting the expensive project. Popular opinion, more directly expressed, was found on all sides of the issue. When required, senators and representatives tapped into local groups and alerted them to the need to deluge DOE with supporting letters. In turn, local opponents, echoing the arguments raised against L reactor, sought to prevent action.

Over the period from 1982 to 1984, the Department of Energy received and responded to hundreds of letters and postcards from concerned individuals and organizations in Idaho. As in the L restart controversy, opinions ran the gamut from fervent support to intense opposition. The organizations included the Snake River Alliance, the Groundwater Alliance, church groups, chambers of commerce, and groups of students. Individuals complained about possible pollution of the aquifer, about contributing to the arms race, about despoiling the scenic countryside, and about the nonparticipatory nature of the decision process. Other individuals insisted that the Idaho site was ideal because of the experienced local labor supply and active support for things nuclear in the area; many feared for the impact on local business if employment declined. Mayors, city councils, and state legislators added their support.

In general, the DOE letter-response system worked promptly, using newly acquired word-processing equipment to compile standard paragraphs into letters that answered, point by point, the individually varied letters of support or opposition. DOE replied to many letters within less than 10 days; considering the volume and variety of the correspondence, the effort was both courteous and remarkable. Letters that filed Freedom of Information requests, in contrast, rarely received prompt action, with the more massive requests from organizations encountering delays that lasted up to several years. Surprisingly, some replies to members of Congress, because of the lengthy process of securing internal concurrences within the department, took much longer than replies to individual citizens.[45] Yet the systematic and organized approach may have created the impression that the letters from the general public had more influence or were given more consideration at a high level than was the case.

On 11 May 1984, Hodel requested congressional approval to reprogram $17.5 million to conduct further studies for an NPR in accordance with the National Environmental Policy Act process.[46] A month later, on 18 June 1984, Representative Price reiterated his earlier objections concerning the possible transfer of DOE funds for that purpose. Although Price did not object to the performance of the studies, he said that accepting continued study did not ensure future congressional approval of the NPR program. Price claimed that the need for an NPR was ill defined and that the size, type, and location of the proposed plant was undetermined; therefore, he claimed, any full EIS study would be wasteful and unproductive; he implied his consent to the technical studies.[47]

Although both the Glennan report and Hodel favored a heavy-water-cooled and -moderated plant, the NPR project office did not drop the Hanford N design from among the proposed conceptual designs. On 16 August 1984, the project support office released a contingency plan for light-water-cooled graphite-moderated technology.[48] Senator Tower's compromise of simultaneous investigation of multiple sites and multiple technologies was in effect; the technology list expanded from two to three.

In the midst of political wrangling over technology and site, nothing had happened to bring an NPR closer to reality. The basic concern by the defense establishment remained alive. On 28 December 1984, Secretary of Defense Casper Weinberger wrote to National Security Advisor Robert McFarlane to suggest that increased special nuclear materials production required an NPR, reiterating Harold Brown's concern about tritium assurance made 4 years earlier. Secretary Weinberger emphasized the need for explicit executive direction to ensure adequate supplies of nuclear materials for future needs.[49]

Despite Weinberger's urging, Hodel was unable to cut through the political deadlock. His resolution and response to the Defense Department request for an assurance regarding nuclear material constituted an admission that little could be done. On 6 February 1985, the last day of his service as secretary of energy before moving on to become secretary of interior, Hodel approved a "reactor production assurance strategy" which recognized a potential delay in NPR acquisition but which asserted that there were "no known near term life-limiting" mechanisms at the Savannah River reactors. On the other hand, N at Hanford was deemed "vulnerable to aging," and the strategy called for a new production reactor to be built by the turn of the century. Barely concealed in the "strategy" was the judgment that the Department of Energy must struggle along with the old reactors for the near future.[50]

Hodel's successor, John Herrington, coming in with the second Reagan administration in 1985, sought to move the production reactor decision along. Herrington continued to seek objective outside analyses that could depoliticize the decision process and allow for a firm choice of technology and site. The extended evaluation process that had proceeded throughout the first Reagan administration, although giving evidence of concern at forthcoming erosion of production reactor capacity, was far less expensive than actually building a new production reactor, variously estimated to cost in the range of $4 billion to $6 billion. One consequence of the protracted technopolitical dispute over new production reactor capacity was an appearance of concern for defense preparedness without actual expenditure of the massive funding required to achieve the preparedness. This technique of walking loudly and carrying a small stick resembled the effort mounted through the Strategic Defense Initiative, in which paper plans and publicity about notional devices may have been as useful in diplomacy as the expenditure of funds on actual, but much more expensive devices.

Chernobyl, N Reactor, and Congress Again

The effort to get along on the surviving old production reactors received a setback when, on 26 April 1986, Unit 4 of the Soviet Union's Chernobyl Nuclear Power Station, an RBMK-1000 graphite-moderated, water-cooled reactor, was destroyed in the world's worst nuclear accident to date. In response to heightened fears, Herrington requested a study by the National Academy of Sciences (NAS) and the National Academy of Engineering (NAE) to assess all the DOE reactors capable of operating above 20 MW(t). NAS produced two studies, one focusing on the existing four production reactors, the other on smaller experimental and testing reactors.[51] Of all the reactors in the United States, N reactor bore the most similarity to Chernobyl, in that it was the only remaining large-scale graphite-moderated reactor in the United States, even though it relied on pressurized rather than boiling water for coolant.

By the mid-1980s, alarming reports of birth defects, cancer deaths, and unexplained illnesses in the area to the east of Hanford led several investigative reporters to focus on the concerns of the downwinders. As previously classified reports of intentional and unintentional radioactive releases over the years came out during these investigations, the barrage of news stories, lawsuits, and revelations of prior coverups mounted. These developments revealed that neither DOE nor the operating contractors could continue to rely on support or mute acceptance from local residents. New local organizations reflected the alignments over the issues. The Hanford Education Action League (HEAL) organized in 1984 to gather more information and to work for closure of Hanford's remaining reactors and processing facilities. Bolstered by the efforts of journalists, freelance researchers, and pastors of churches, the organization soon grew in sophistication and research skills. Another group, the Hanford Patrol, collected technical information revealing incidents of radioactive pollution. A reporter for the Spokane *Spokesman-Review*, Karen Dorn Steele, published a series of articles revealing specific downwinder cases and release incidents that might have contributed to their maladies and deaths. In self-defense, a group of engineers formed the Hanford Family, an organization devoted to offsetting the negative publicity and protecting Hanford from its detractors. In April 1985, HEAL published a white paper which revealed that in 1959, Hanford had released over 20 curies of radioactivity every day of operation, more than had been released in the whole Three Mile Island accident.[52]

While the NAS studies regarding DOE's reactors were in preparation, protest against N reactor's continued operation flowed in from a variety of sources. The Nez Perce Indian tribe in the state of Washington demanded its immediate closure. Congressman James Weaver of Oregon introduced a resolution in Congress asking DOE to keep N reactor closed pending investigations by the General Accounting Office and others.[53] Internally, DOE's Office of Environment, Safety, and Health conducted a design review and a technical safety appraisal of N reactor, suggesting a variety of safety improvements.[54] DOE announced a planned set of accelerated changes in N reactor design in response to the appraisal.

On 12 December 1986, a Herrington-appointed group, the Roddis Panel, completed its evaluation of the Savannah River reactors and the WPPSS Nuclear Project Unit 1 (WNP-1) reactor. The chair of the panel, Louis Roddis, Jr., was a product of the Rickover network, had served in the Naval Reactor Division of the AEC, then as deputy director of the Reactor Development Division of the AEC in the late 1950s, and as president of Consolidated Edison of New York in the early 1970s. Roddis had chaired the Energy Research Advisory Board from 1981 to 1984, and his selection to evaluate NPR issues was an indication of the continued search for prestigious and objective technical policy input.

The Roddis panel concluded that the aging production reactors at Savannah River were not reliable for defense needs but if upgraded, could operate for 5 additional years. The panel, as well as GAO investigators, recommended a permanent shutdown of the Hanford N reactor. N reactor had no containment vessel and would never have passed NRC licensing requirements had these requirements been applied to DOE reactors. The Roddis panel pointed out that the reactor did not even have a hydrogen control or hydrogen monitoring system, which had been present in the Chernobyl system.[55]

In response to national and local public outcry, N reactor was put on stand-down in January 1987.[56] Later in 1987, NAS and NAE issued the DOE-requested report, entitled *Safety Issues at the Defense Production Reactors*. This report, which focused on all four production reactors, was highly critical of the Department of Energy and its reactor management. NAS indicated that DOE had relied "almost entirely" on its contractors to identify safety concerns and that the federal government had not "realistically addressed the aging of the defense production reactors." Safety oversight, according to NAS, had become "ingrown and largely outside the scrutiny of the public." Planning for new production reactors should accelerate, the report concluded.[57]

The NAS-NAE study, more circumspect about N reactor than some of the others that came on the heels of the Chernobyl disaster, pointed out a number of significant differences between the Soviet RBMK design and N reactor design. Among these were the fact that N reactor hydraulic control rods could enter the reactor in 2 seconds, whereas those at Chernobyl were gravity-driven and took 20 seconds to fall in place. Chernobyl used boiling water as a coolant, rather than pressurized water as at N; boiling water could create voids, causing potential unstable power excursions. Chernobyl did not have the backup boron–carbide ball safety system that N used. The confinement system at N reactor allowed release of excessive pressure through filtered pathways to the environment; the containment system at Chernobyl provided no pressure relief and simply ruptured. Nevertheless, NAS concurred that N reactor should stay closed.[58]

The closing of N reactor at Hanford and growing concerns about the long-term leaks and newly diagnosed intergranular stress corrosion cracking in the C reactor vessel at Savannah River and its subsequent closing in 1987 provided a stimulus to the NPR effort, which had been submerged in studies since 1980. Less than 2 years after Hodel had announced that there were no apparent "life-limiting" factors in the Savannah River production reactors and had regarded that as sufficient "assurance" of productive capacity for the Defense Department, C reactor there had been closed for safety reasons. N reactor, which Hodel had admitted was vulnerable, had also closed in the wake of Chernobyl. In February 1987, DOE's deputy director of Defense Programs Charles Halsted notified Under Secretary Joseph Salgado of his concern over meeting the stockpile memorandum tritium requirements with the elimination of N and C reactors as reliable producers.[59] Halsted recommended immediate action on the NPR, although he did not specify site or technology choice.

Through 1986 and 1987, Westinghouse Hanford Corporation, the contractor in charge of operating N reactor, tried to forestall the storm of post-Chernobyl criticisms by engaging in a vigorous program of safety enhancements. Meanwhile, the Tri-Cities Development Council (now operating as "TRIDEC"), the congressional delegation from Washington, and the DOE Richland Operations Office all worked to preserve employment at Hanford, much as they had during the 1960s. Sen. Dan Evans, Rep. Sid Morrison, and TRIDEC presented materials suggesting that plans should proceed to complete WNP-1 and its conversion to tritium and plutonium production. TRIDEC funded a legal study that was submitted to Congress, examining exactly how the ownership and jurisdiction over WNP-1 could be shifted to DOE. After a dispute over the proper roles of the Richland Operations Office and Hanford contractors in providing material and briefings to political representatives and others outside the department, Under Secretary Salgado ordered that further draft materials on WNP-1 not be circulated. Congressional objections to the use of federal money on the part of the contractor in lobbying Congress had prevailed.[60]

Herrington still sought an objective report on which to base a nonpolitical resolution of the issue of reactor site and technology choice. On 7 January 1988, he asked the Energy Research Advisory Board to review and evaluate four reactor technologies for NPR capacity.[61] On 28 January 1988, the board established a site evaluation team to develop criteria and evaluate DOE-owned sites for new production reactor capacity, and on 25 February 1988, Herrington made an interim report to Congress on NPR selection strategy activities under way by the board and the site team. Herrington requested that the board's criteria for technology selection include duality, which he defined in a new, more attractive fashion. Whereas earlier, "duality" had implied that the new reactor was to represent redundancy with any surviving older reactor or reactors, in 1988 Herrington suggested that Energy Research Advisory Board identify two technologies and two sites for the new production reactors in its assessment.[62] The concept was politically attractive, for it could foster the alignment of congressional delegations from the two preferred sites to support the project, possibly breaking the deadlock. Herrington informed the pertinent congressional committees about progress, explaining his concept of duality, activities regarding initial procurement, and plans for proceeding with the National Environmental Policy Act process.

Through 1988, the remaining three Savannah River reactors were shut down out of concern for safety: K on 10 April, L on 23 June, and P on 17 August. An attempt in August 1988 to restart P reactor after reinstallation of seismic bracing was foiled by the presence of helium-3, a tritium decay product that had been unintentionally produced from the deuterium moderator. Since helium-3 acts as a neutron absorber, the reactor did not start at the removal of the usual number of control rod equivalents. Operators removed an extra 60 rods before deciding to review the problem and search for its cause, a procedure roundly criticized in later analyses, particularly by former Nuclear Regulatory Commission chairman John Ahearne.[63]

Building an NPR required a major commitment of funding from Congress; even to proceed with conceptual design work and selection of a technology required hundreds of millions of dollars. On a much larger scale, the Department of Energy faced severe problems regarding cleanup of radioactive and hazardous wastes that had accumulated at the weapons complex sites for decades. Initial estimates that cleanup costs might exceed $100 billion were daunting. Over the period October to December 1988, relatively quiet hearings of the Armed Services Committees of both houses of Congress on production reactor issues became front-page news across the United States, partly because of attention given to problems of radioactive waste. News stories focused not only on the need for cleanup, but also on safety at production reactors and on the cost of replacing those reactors.

Secretary Herrington took the unusual and bold step of discussing these issues publicly, in contrast to the well-established AEC-DOE tradition of working behind closed doors, especially on issues as potentially disturbing to the public as massive waste and high future expenditures. In October, for example, Herrington met with the editorial boards of the *New York Times* and the *Wall Street Journal* and appeared on major network news interview shows, including NBC's "Today" show, CBS's "This Morning," and the NBC "Nightly News." In addition, the department was forthcoming in releasing details of 30 serious reactor incidents over the years at Savannah River, which the press soon dubbed "the dirty thirty." The media feeding frenzy that began in early October may have been the result of the *New York Times*' initiating an old-fashioned journalistic crusade in the muckraking tradition. To an extent, the media coverage seemed to derive strength from a press habit of defining as newsworthy those items that two or three leading papers chose as front-page material, the same pattern that accounted for short periods of intense press interest in other single stories. Furthermore, Secretary Herrington's willingness to be open-handed about departmental needs when faced with a difficult budget argument provided good copy. However, behind the media attention lay a seriously eroded safety regime, hidden by a distorting culture of secrecy. Reporters like Keith Schneider of the *New York Times* had performed a valuable service in exposing the lack of training and the incompetence of reactor personnel at the surviving production reactors.[64]

Through November 1988, DOE held "scoping" meetings in Idaho and South Carolina to obtain public reactions to expansion of reactors at the two locations. In December, DOE sent to Congress the *United States Department of Energy Nuclear Weapons Complex Modernization Report*, (known, more conveniently, as the 2010 Report), recommending the construction of new production reactor capacity as an aspect of upgrading the entire weapons complex over the next 15 to 20 years.[65]

Decisions

With the closure of the last of the Savannah River reactors, the Senate Armed Services Committee considered a series of options to ensure a supply of tritium, including reconfiguration of weapons, recovery of tritium from low-priority weapons, restart of one or more Savannah River reactors, and even a restart and low-power operation of N reactor, converted for tritium production only. A new production reactor, most of the committee members agreed, was required to ensure against a shortfall by the turn of the century.[66]

Meanwhile, informed advocates of disarmament noted that, without tritium production, nature itself would generate disarmament. In a well-documented book, J. Carson Mark, Paul Leventhal, and others argued that the "Tritium factor"—the 5.5% per year decay rate of the strategic isotope—would start the United States on the path of a declining weapons stockpile. If the USSR agreed to halt tritium production, the decline in weapons would proceed at a rate even higher than that proposed under the agenda for the Strategic Arms Reduction Treaty (START) under discussion with the Soviet Union. Simply not replacing production reactors would automatically generate disarmament.[67]

That approach did not prevail, however. In January 1989, DOE submitted its fiscal year 1990 budget request to Congress and included $303.5 million for NPR work; the department also released a declassified version of the 2010 Report, further publicizing the need for weapons complex modernization.[68] On 19 January Secretary Herrington sent a set of plans to Congress, "Actions to Shorten New Production Reactors Schedules." The second Reagan administration thus ended with recommendations to begin work on the reactors that had been discussed for 8 years.

The 1989 plans reflected the 1988 Energy Research Advisory Board report and called for two reactor developments: one to produce 100% of the tritium requirement and a cluster of reactors based on an innovative and safe design, a gas-cooled, graphite-moderated model, which could produce 50% of the requirement. The gas-cooled units, technically most efficient on the smaller scale, would be built in a group of reactor "modules," allowing support facilities to service more than one reactor. Design elements worked out on these reactors might serve as models for other applications, such as power generation. Yet to implement the plans required continued congressional support. Herrington's legacy to his successor was a strenuous effort to cut through the gridlock, yet no design had been chosen, no contractor committed, no final site selected and approved through the EIS process. The plan made sense, but no firm decisions had been taken.

Several factors over the early and mid-1980s had immobilized the nation's ability to make a decision to rebuild its nuclear weapons producing capacity. Decisions once reached by General Groves in consultation with a selected group of specialists now were open to discussion in Congress and in the public. Throughout the nation, antinuclear groups had grown in experience and in organizing ability. Journalists writing for daily newspapers across the nation criticized the Department of Energy for its emphasis on production over safety and concern for radioactive pollution and improper waste handling. While most Americans knew little of production reactors, those who stood to lose their livelihoods at Hanford or Savannah River had effective political voices. At Hanford, politicians worked closely with local leaders and with technical specialists; yet INEL and Savannah River had effective spokesmen in Senators McClure and Thurmond. In order to get action, the secretary of energy needed to make an unbiased choice among potential sites and technology, a choice that could not be instantly criticized and blocked by charges that it was hasty, ill-informed, technically incorrect, unduly influenced by special interests, or inconsiderate of impacts.

9 • Managing Nuclear Options

Energy secretary John Herrington left his successor a daunting task. Like Herrington, the new secretary had to operate a sprawling, multibillion-dollar department with inherited responsibility for facilities from the Manhattan Engineer District and the Atomic Energy Commission, facilities which grew increasingly unsafe year by year. He had to operate amidst growing public and congressional awareness of the vast environmental hazard generated in the nuclear complex. The new secretary appointed by President George Bush, retired admiral James Watkins, approached the job with a determination to make changes. Secretary Watkins confronted the impending tritium shortfall that threatened the viability of the nuclear arsenal, but in order to deal with it, he had to finesse the politics of the production reactor design decision.

To fully remove the issue of how to maintain production reactor capacity from the political forum was not possible. Senators James McClure of Idaho and Mark Hatfield of Oregon warned Watkins, early in his term, of what he faced. They pointed to the decade of studies of the tritium production issue and decried the fact that "several of the options which were rejected in this decade of study and debate are once again being touted by their political, technical and economic beneficiaries.... We have continued to see articles, press releases and open politicking for these alternatives." Members of Congress were being lobbied to convert a partially completed light-water Washington Nuclear Power (WNP) reactor into a production reactor. Such a conversion, said McClure and Hatfield, would "cast a long, ominous shadow over this country's commitment to nuclear nonproliferation." Similarly, advocates of restarting N reactor were hoping to reinvigorate the old graphite-moderated water-cooled approach, despite the 1988 commitment to restrict the approach to two technologies, heavy-water and high-temperature gas-cooled. "We do not have the luxury of another decade of committees, panels and studies," concluded Hatfield and McClure. For their part, defenders of Hanford, like Congressman Sid Morrison, claimed that the exclusion of WNP-1 "reflects a dramatic disregard for either project cost or assurance of timely completion," and such rejection by Hatfield and McClure represented "politics as usual."[1]

The approach and style that Watkins brought to the overall management of DOE reflected quite a departure from those of his predecessors. He attempted to establish a way of dealing with the tough technical decisions over site, contractor, and conceptual design in a goal-oriented manner like that of Leslie Groves. A detailed examination of Watkins' administration shows how he set out to reach such decisions promptly and objectively in the altered political environment of the 1990s. During his administration, every decision and action drew the attention not only of local newspapers in South Carolina, Georgia, Idaho, Washington, and Oregon, but also of the *Washington Post* and the *New York Times*. He conducted his attempts to make changes under the spotlight of full news coverage.

The Management Culture

At his confirmation hearings before the Senate in February 1989, Watkins explicitly said that he intended to change the "existing culture" at the Department of Energy. While many observers in the press and in professional nuclear circles agreed that the culture needed change, his statement was open to various interpretations.

Watkins' acceptance speech and his actions reflected the concept of "corporate culture" developed by management theorists and practicing managers earlier in the 1980s. Between 1979 and 1982, the concept of "corporate culture" had entered the day-to-day vocabulary of managers. Several best-selling management books, including *In Search of Excellence* by Thomas J. Peters and Robert H. Waterman, Jr., further spread the concept of attempting to understand a corporation by examining its cultural behavior.[2]

These and other writers on management used language drawn from sociology and anthropology to suggest that each corporation developed cultural norms that shaped its effectiveness. In this view, some corporations had "strong" cultures, rich with customs, legends, and behavioral expectations that reinforced the corporation's mission. Peters and Waterman argued that some strong corporate cultures were dysfunctional, often committed to out-of-date approaches, while others were well suited to the modern marketplace and the world of international technological competition. Although most corporate culture theorists agreed that a company's cultural pattern determined its behavior, they sharply disagreed over whether these patterns could be changed.

In any case, none of the most popular works on corporate culture in the private sector delved into the existence or nature of a culture at specific federal agencies, providing little in the way of direct guidance to Watkins, had he sought it. Most of the management literature used the social science term "culture," with its implications of a broad meaning, to refer to a narrow set of interacting corporate practices which affected management. For many managers and administrators the term "culture" was a contemporary way of describing management style, and it was in this narrow sense that Watkins used the term.[3]

Watkins' verbal attack on the culture in the weapons complex and later in the whole Department of Energy aroused the expectations of observers, for many read into his comments a broader intent. Some antinuclear and environmentalist critics of the DOE saw their own views as part of a broader cultural transformation in the nation. For example, Lewis Shaw, a South Carolina environmental official, claimed that DOE "got caught up in a time warp" in the late 1970s and had fallen 20 years behind the rest of the nation.[4] By the early 1980s, a wide gulf had in fact developed between the cultural values that had gone into the creation of the weapons complex and the values of the broader society outside the fences.[5]

Both within and outside DOE, critics questioned exactly what Watkins meant by the departmental culture and speculated about what aspects he planned to change. Recent concerns expressed by Herrington, members of Congress, and the press about previously unpublicized environmental issues at the weapons complex raised the expectations of some groups that the new secretary meant not only to address these environmental concerns but also the tradition of confidentiality which had limited public information about those problems for decades. Items in the press suggested that both environmentalists and antinuclear activists had hopes that Watkins' statements heralded a shift away from reliance on nuclear weapons and toward openness. One writer, searching for evidence of a new style, spoke of "Radio Free

Watkins."[6] Environmental groups later issued annual "report cards" on Watkins, claiming he failed to meet his own standards, or their expectations, on public access to information and protection of whistle-blowers.[7]

But when Watkins spoke of the departmental culture that he wished to change, he employed the more specific language current among managers rather than the broader concept of national culture employed by environmental critics and journalists, a fact demonstrated by his actions, which focused on management weaknesses at DOE. He intended to strengthen the weapons complex and its technology, not diminish it. Watkins' approaches to administrative problems can be put in perspective through a glance at his background in management as conducted in the modern Navy.

The Navy's Management Culture

Admiral James Watkins' career as a naval officer spanned more than three decades, as he rose through the nuclear Navy under Hyman Rickover to the highest office the Navy offered, chief of naval operations. Naval reforms in management styles reshaped the Navy over those years. In particular, the idiosyncratic methods employed by Rickover and the more widely emulated management-by-objective style implemented in the Polaris Special Project Office under Adm. William F. Raborn and Adm. Levering Smith affected the Navy's management of its large-scale research and development efforts in the 1950s and 1960s. Further changes resulted from Secretary Robert McNamara's systematic reform of the whole Department of Defense in the 1960s, which incorporated management-by-objective methods, a demand for excellence, and the systems methods that had characterized the pioneering work of Rickover and Raborn. Watkins, as an officer selected by Rickover, served as a nuclear submarine commander before moving up in the Navy hierarchy.[8]

Rickover's style had been intensely personal over the period from the 1950s through 1982 during which he directed the Navy's nuclear propulsion effort. He was skeptical of respect for rank alone, demanding intellect and performance as well. To design and build nuclear submarines and surface vessels, he believed he needed an independent command, with guaranteed funding and minimal interference from naval administrators who put priorities on cost instead of quality and who traditionally rewarded rank instead of achievement. By working directly with Congress, Rickover had secured a degree of independence from the conventional system of naval procurement. He selected individuals for his program on the basis of a demanding standard and then held them to high levels of performance through a combination of ruthless drive and biting sarcasm. He established his own training schools, which turned out hundreds of nuclear engineers, many of whom moved into positions in government, manufacturing, and utilities after retirement from the Navy. Despite the fact that his style won him many critics and enemies, Rickover achieved what he set out to do: he established a standard for nuclear safety and quality work and an esprit de corps which, at the program level, represented an intense and effective culture in itself. The result was an elite leg of America's triad of strategic defense, consisting of land-based missiles, aircraft, and nuclear submarine–launched weapons. Rickover also produced a national network of former naval officers experienced in nuclear matters and dedicated to the ideals he had espoused. Veterans of his program still recount anecdotes and experiences from their days under Rickover, reflecting the pattern of symbolic legend building. While his demanding search for excellence was legendary, his idiosyncratic management style was difficult to emulate.[9]

Through the same period, the Navy also modernized its internal market arrangements, which had a long history in the service's bureaus. Headquarters systems program officers "purchased" research and development from Navy-owned and -operated laboratories, testing facilities, and experimental stations. The system of Naval Industrial Funding formally set up the Navy laboratories, research and development centers, and other supplying facilities on a quasi-competitive basis in which they would secure much of their funding from "customers" at the systems command levels, rather than through direct annual appropriations to the facilities.

Although the nuclear propulsion program under Rickover and the Polaris missile Special Projects Office under Raborn were unique, elements of the demand for excellence as well as the high standards they set came to characterize much of Navy purchasing. Since systems program officers, each looking out for their own projects, had money to dispense and a wide variety of laboratories and other facilities to choose from, they could shop around for the best product inside the Navy's own establishment. The result was a sometimes highly competitive struggle for funding, recognition, and projects among the Navy's research and development facilities. Despite notorious overruns and program cancellations in the procurement of ships and aircraft from private manufacturers, several major innovations and many minor improvements in weapons, ships, communications equipment, and a wide variety of technical systems and subsystems came out of the Navy's own in-house labs and centers.[10]

The Navy's reliance on government-owned, government-operated facilities for much naval research and development did not go unchallenged during the Cold War era. Department of Defense and naval reformers sought ways to foster innovation and independence inside civil service structures and under the command of uniformed officers. The Navy had been racked by decades of dispute and reorganization, centering on debates over the most viable size of laboratory units, over intellectual freedom and competitive pay, over the need to coordinate the needs of users with the ideas of producers, and over repeated efforts to downsize government payrolls and privatize government capabilities.[11]

During his tenure as chief of naval operations in the 1980s, Watkins worked with Secretary of Defense John F. Lehman, Jr., to tighten further the Navy's system of procurement from the private sector, trying to ensure adherence to high quality and competition and imposing fixed-price contracts in some situations in which cost-reimbursement, and hence expandable, contracts had prevailed. In this and other ways, Watkins had reflected the Navy's experience in controlling private contractors through the internal establishment and enforcement of high standards.[12]

Although the Navy's approach to research and development could demonstrate success, this approach was difficult to imitate or to export to other government agencies. Rooted in more than a century of a structured relationship between suppliers in Navy bureaus and users in the fleet and created through decades of sometimes contentious reforms, the Navy's modern systems approach was built on military command and military policymaking. By contrast, the DOE weapons complex, with civilian management roots in the Atomic Energy Commission of the 1940s, had evolved with an entirely different relationship between military end-users and the private-sector research and development institutions that created the products. It was clearly impossible to scrap the entire existing DOE institutional structure; nevertheless, values and goals derived from the Navy approach might be applicable in an effort to improve DOE performance.

When Watkins sought to implement reforms in the Department of Energy, the vocabulary he employed consciously invoked the overtones of a naval background, with its emphasis on safety, engineering excellence, and accountability. Watkins believed in the virtues of the Navy's strong policies of expecting, demanding, and getting performance out of private contractors through tough and informed in-house managers with strong technical know-how. Watkins explicitly emphasized his debt to aspects of the Rickover example. However, such an approach implied that he was unaware of or overlooked the fact that the Rickover values and goals—the Rickover "effect"—had already permeated the nuclear engineering community and much of the weapons complex. Former naval reactor personnel and former Rickover officers were scattered throughout the nuclear weapons complex, both in federal and contractor positions. They and other staff were justified in resenting the implication that they did not already pursue technical excellence, safety, and professional quality, sometimes quite consciously in the Rickover tradition.[13]

The Old Culture at DOE

Some of the very characteristics of the DOE culture that Congress, the press, and state officials criticized were to some extent typical of military technology enterprises in the Navy as well, and were not necessarily targets of Watkins' intended reforms. From the older AEC culture, DOE had inherited the system of remote siting, fenced-in compounds, the habit and practice of secrecy, and the routine control of information that could flow to the media. These traits—which had emerged from the uneasy blending of industrial, military, and academic elements under Gen. Leslie Groves—continued to permeate the weapons complex despite the intent of the 1946 Atomic Energy Act to place nuclear weapons manufacture under civilian control. During the 1980s such practices had come under considerable criticism from Congress and the press as "civilian control" was increasingly defined to mean open and public participation in decisions. These criticisms received a form of official endorsement from the hard-hitting post-Chernobyl National Academy of Sciences study of 1987.

In DOE, as in the military, mistakes when made were not publicized but dealt with quietly. Issues such as risk, worker safety, and pollution were taken seriously and enforced through internal organizations behind the wall of secrecy. In DOE, at the heart of strategic material production issues, crucial information for informed opinions and decisions remained hidden in darkness. Only a limited circle of decision makers had access to and the "need to know" the specific size of the stockpile of strategic materials, the quantities of tritium produced and anticipated, and the quantitative impact of continued nonproduction. Outsiders and, presumably, Soviet intelligence officers and planners could make informed guesses, but details were not public. In all of these ways, DOE and the military services were not so different.

Beyond this, even unclassified weapons complex information and data with far less strategic importance were habitually not widely known or disseminated. For a few months at the end of his tenure, Secretary Herrington had stepped away from the traditional culture of secrecy for a specific political purpose when he openly discussed the problems of cleanup and modernization in the wider forum, as a tactic to raise congressional willingness to provide funding.

One major contributing factor to the Department of Energy's problems in the 1980s was the sheer size of the weapons complex and the administrative difficulty inherent in overseeing it. The effort by both Carter and Reagan to cap bureaucratic growth had weakened the ability of technical government employees to oversee the work of contractors.[14] In this context, DOE administrators found it politically difficult to increase the number of department personnel. Consequently, they continued to rely on the widespread network of contractor-operated facilities and contractor-performed work to meet the demands of an expanded weapons program. The longstanding tendency of the laboratories and production facilities to be locally directed, the origins of which could be traced to the tensions between the field and headquarters under Groves and then under Lilienthal, was sharpened, not reduced, during the Carter and Reagan era. One consequence of diminished oversight was sometimes collusive arrangements between DOE field office staff and local contractor staff, an issue which surfaced as front-page news during Watkins' administration.[15]

The reactor sites all continued to operate under administrative contracts modeled on those first established by the Office of Scientific Research and Development and General Groves and then reissued by the Atomic Energy Commission. Contractor-operated facilities, particularly Savannah River and Hanford, operated as huge employers of several thousand persons each, directed by relatively small headquarters offices and token "area offices" of federal employees at the sites. Unlike the Navy, DOE had difficulty maintaining an internal elite corps of technically proficient government-employed experts who could effectively monitor the work, relying from the beginning on both academic and corporate contractors to perform that supervisory work.[16]

The sheer size of the contractor-operated field facilities further hampered headquarters' ability to maintain accountability. By the 1960s the AEC complex had about 7,000 federal employees and 170,000 contract and academic employees, a ratio of about 1 to 25.[17] The volume of paper and the vast amounts of data produced by the national laboratories and production operations outpaced the capacity of the relatively small headquarters and area office staffs to manage. The department's own inspector general pointed out this problem, as did congressional critics such as Mike Synar of Oklahoma, who claimed that the weapons complex was "out of control."[18] In Washington, by the 1980s even headquarters functions came to be handled by "support contractors." In general, DOE was not able to secure adequate funds for maintenance, expansion, rebuilding, or improvement of its physical facilities or to increase the staff involved in oversight of the contracts.[19]

During the 1980s contractor operations boomed, while the government-owned facilities and staff tended to be neglected. The central administration rejected repeated requests from the field for capital improvement, maintenance budget expansion, or more federal specialists to oversee the contractors. What some outsiders criticized as a "culture" of neglect or complacency within DOE derived from the fact that headquarters had few alternatives to accepting contractors' technical information.[20]

Many of the Navy laboratories, by contrast to DOE's, were staffed not by contractors but by small cadres of naval officers and enlisted men and larger numbers of civilian naval employees directly on the Navy's payroll. However, nowhere in his remarks did Secretary Watkins indicate that he wished to eliminate the system by which major corporations contracted with DOE to operate the weapons complex sites. Despite the criticisms of the fundamental institutional structure of the GOCOs, Watkins did not set out to undo those contracts or restructure that whole system. Rather, he attempted to improve the system's quality and its performance, values, and expectations. Some of the reforms implemented by Watkins tightened and altered the way in which DOE managed contracts. He sought to employ the technical firms in ever more efficient and accountable ways and to insist that DOE's own supervising program officers take responsibility for ensuring that the contracts were properly fulfilled. In effect, Watkins hoped that the department's program officers could begin to play a role similar to the systems program officers in the Navy's funding arrangement for its research and development, while still relying on the basic GOCO structure.

Similarly, he never suggested an attack on the system of classification of information and the maintenance of safeguards and security, which outside critics such as Ralph Nader had viewed as characteristic of the AEC culture and which a rising chorus of critics complained about by the 1980s. Indeed,

Watkins' administration moved to strengthen that system, requiring that security rules be followed even more closely at both field operations and headquarters.

For such reasons, it was incorrect to view Watkins' reforms at DOE simply as part of the broader national cultural shift away from World War II consensus values, the values that had shaped the early Manhattan Engineer District. Watkins tried to improve operations at the department, but he did not try to move the agency from a technological and authority-based system in the direction of a humanistic, "smaller is better," nonnuclear world. When journalists and environmental activists heard of a cultural revolution, some appeared to believe that the age of high technology and decision making by experts was about to give way to a wave of public decision making, especially on all matters affecting the environment and open disclosure; perhaps it could even provide an end to nuclear technology itself. But Watkins sought to implement an age of accountability, not the Age of Aquarius.

Management Reforms at DOE

The changes and reforms that Watkins implemented, as well as his widely publicized statements, show exactly what sort of cultural change he sought. The procedures he set up to seek excellence and accountability gradually affected parts of the sprawling DOE establishment under his administration. Yet inside the Department of Energy weapons complex and among nuclear engineering professionals in contractor organizations, some of Watkins' statements and his particular reforms were greeted as if intended only for public or congressional effect. In truth, he took actions that reflected his public stance and affected the internal structure as well.

Watkins made his intentions clear to DOE personnel by issuing statements as "Secretary of Energy Notices," as well as through a series of press releases, on the need for change at DOE. In addition, he appointed individuals who reflected the attitudes and behavior he looked for, and he also enacted specific reforms intended to address the problem of accountability.

Watkins's appointments during his first months of office were part of this effort to implement change in the management culture. While incoming departmental secretaries normally began their term of office with a new cadre of upper-echelon officials, his own appointments placed an emphasis on selection of people with both technical and administrative, not simply managerial, experience. He attempted to attract highly qualified individuals from industry by seeking to change the "revolving-door" rules that prevented federal officials from moving from the private sector to government and back again. Furthermore, he sought approval to increase salaries of top scientific and technical personnel in order to make federal employment more competitive for highly qualified scientists and engineers. Some outsiders had hoped that the cultural change would be represented by the recruitment not of experienced science and technology managers but of policymakers with a reputation built on opposing development, particularly nuclear development.[21]

Admiral Watkins implemented a host of measures to bring about accountability, to instill a "safety culture," and to improve relationships between the weapons complex facilities and local governments. He established independent "Tiger Teams" to evaluate the major centers and tightened both safety and security regulations.[22]

The changes generated some crises along the way. For example, public disclosures by the DOE's inspector general of inappropriate transfers of funding from construction accounts to operating expenses by Savannah River officials resulted in both a short national scandal and replacement of the officials. At headquarters and at Savannah River, the misuse of funds was attributed to bureaucratic inertia and to the persistence of the old culture.[23] Henson Moore, Watkins's deputy secretary, complained that the field office reflected the "same kind of culture and how this place has been run since the day it opened its doors." He was "furious" over the crisis.[24] Watkins replaced the manager of the Savannah River Field Office with a 39-year veteran of the nuclear Navy, Vice Adm. Peter M. Hekman, Jr.[25]

What Watkins had defined as cultural change, and what in fact was an attempt at management reform, shaped the institutional environment in which a serious effort was mounted to settle upon a new production reactor design. Watkins had inherited from Herrington the Office of New Production Reactors (ONPR), created on 1 October 1988,[26] to be devoted to sorting out the design choices. That office had operated through the last months of Herrington's administration with a small staff, mostly carried over from the previous DP-132 office, under the acting directorship of Ron Cochran. On the organization charts of the department, ONPR had a rank equivalent to that of Defense Programs, which managed the whole weapons complex, reflecting the importance attached to the effort by Secretary Herrington. In order to invigorate that office and to move along the production reactor effort, Watkins conducted a search for a director of the office whose background would combine a knowledge of DOE, a technical background, and experience with the Navy, but without any commitment to one or another of the prevailing sites or conceptual designs. Rather than seeking an engineer with direct reactor experience, Watkins looked for someone with administrative and managerial knowledge and experience to build the large team that would be required to sort through the complex issues of policy, contracting, design, technopolitics, personnel, and budget. It took until midsummer of 1989 to find the proper candidate.

New Management at New Production Reactors

On 12 June 1989 Watkins appointed Dr. Dominic J. Monetta as director of the Office of New Production Reactors. It would be Monetta's task to provide personal leadership and bring energetic management to the long-delayed effort to replace the aged reactors. When he took office, all of the production reactors were closed. Both K reactor at Savannah River and N reactor at Hanford were in standby status, and Westinghouse at Savannah River was charged with bringing K reactor up to safety standards for operation.

Monetta would not have any jurisdiction over those existing production facilities, whether closed or on standby, but would concentrate instead on bringing the plans for a new reactor to fruition. He would have the task of assembling and managing a large-scale, complex technical effort on a tight schedule, but without the war-induced urgency and secrecy of the 1940s or the 1950s. He faced a vastly changed political environment from that in which Groves and the early AEC had operated, and like Watkins, he had to operate in the glare of media exposure.

Because ONPR was a new office with an expanded mandate, Monetta was in a strong position to implement a fresh approach to the long-standing issue of selecting an NPR conceptual design, particularly since he could assemble staff from outside the existing department as well as from inside. Monetta's direction of the NPR effort can be seen as the application of modern management procedures to the difficult technological and policy choices facing the weapons complex, and this phase of the production reactor story necessarily focuses on issues of innovative managerial procedure and their application rather than upon more strictly engineering and technical problems. Most of the particular administrative styles and methods of ONPR could be viewed as implementations of Monetta's own ideas of management, drawn from his own professional background.

Monetta held a B.S. in chemical engineering and a doctorate in public administration. He had worked as a civilian chemical engineer at the Naval

Ordnance Station at Indian Head, Maryland; as manager at the ERDA Office of Conservation; and as senior executive in the DOE Office of Fossil Energy. Following his work in the Office of Fossil Energy, he had set up the planning and analysis functions at the Gas Research Institute and had been an independent consultant for energy research and development organizations. Most recently he had served as technical director at the Naval Ordnance Station at Indian Head.[27]

Watkins did not seek an administrator with a doctorate in physics, chemistry, or engineering. Rather, he scoured the federal government for an experienced technical administrator with pertinent undergraduate training and a strong record in administration, and he found Monetta through a review of the resumes of Navy laboratory technical directors. Monetta met the parameters set by Watkins in staffing the post: significant technical background and specific energy experience, a record of senior administrative responsibilities, and close familiarity with the Navy's accountability practices as a former senior civilian technical executive with the Navy. Watkins saw a strength and an advantage in the fact that Monetta had no career linkage to any particular nuclear reactor technology, nor to any of the corporate interests engaged in reactor design, nor to any of the three sites; there would be no suggestion of conflict of interest as he worked through the "downselect" processes. "I was brought in," Monetta said, "particularly because I do not have a site preference or a technological bias." In one sense, the administrators of nuclear weapons facilities, however experienced, would be disqualified because their experience identified them too closely with one or another of the options. Similarly, nuclear engineers or physicists with experience on one type of reactor would be subject to that same charge of technopolitical bias.[28]

Even in some of his role models, Monetta went outside the nuclear establishment. He revered his mentor at Indian Head, Joe L. Browning, an energetic engineer-administrator who had been technical director of the facility in the 1960s. Browning himself had worked under Admirals Raborn and Levering Smith at the Polaris Special Project Office and liked to view his demand for excellence as technical director at Indian Head as part of that Raborn-Smith tradition. Browning prided himself on selecting young engineers to serve on an assistant management board for the purpose of exposing them early to sophisticated management issues and sharpening their administrative potential. Monetta was one of those selected to serve on and eventually chair that board, and his first experience with administration was in this particular institutional culture, which idealized the concept of excellence. Monetta traced the origins of his ideas back through Browning to the Polaris program when explaining his concepts to others.[29]

Monetta personally interviewed every new appointee to ONPR, in the Rickover tradition. Between his appointment in July 1989 and November 1991, Monetta built the office from a staff of less than 12 to one of over 350. He selected staff with backgrounds in technical administration from DOE, from the Navy, from nuclear power utilities, from the Nuclear Regulatory Commission, and from the Tennessee Valley Authority. He regarded them as each coming from a distinct corporate culture, referring to the result as a form of "cultural diversity."[30]

Monetta emphasized accountability and responsibility. In particular, he expected his technical staff to directly manage contracts. When working with the strong-willed operating contractors at the national laboratories, Monetta had a group of specific administrative tools that he described in explicit language. He tried to work with "dedicated cells" and "a single point of contact for the field." By these terms he meant that a particular officer in his organization would be responsible for a single contract and that the contractor would deal directly with that representative. Such an arrangement prevented the contractor from playing one administrator against another. The point of contact in the contracting organization had to be the "administrative head of the unit" performing the work. In the field, Monetta expected to be represented by "dedicated consolidated offices" and to be allocated "whole man years." By this procedure, he sought to avoid evasion of responsibility through the argument that the work could not be done because of the claims of other DOE programs on an individual's time. Monetta's methods of seeking accountability among contractors reflected models established by Rickover.

Monetta also expected contractor organizations to maintain offices in Washington, D.C., so that meetings could be held and contacts made without excessive travel on the part of his overworked federal staff. As might be expected, his methods sometimes irritated long-term DOE staff members in established offices and some contractors who were used to a less demanding style and pace; others found the new approach refreshing.

Monetta described his administrative guidelines in a rapid-fire vocabulary derived from his combination of engineering and management background. He wanted the ONPR subculture to be "results-oriented" and what he called "oriented to short time constants." He compared that concept to "running a whole marathon in one-hundred-yard dashes." Monetta characterized the old DOE culture as putting the blame on the contractor for errors or shortfalls; he characterized the new culture as placing the responsibility upon the DOE manager, the contracting officer's technical representative. He selected technical representatives who were well informed about procurement regulations and had background in the particular scientific and engineering specialties of the contracts they administered, a pattern very similar to that in the Navy and in the Rickover tradition.[31]

Reflecting the social science orientation of his management degree, Monetta tried to influence the growth of the informal organization. He attempted to build a sense of team through establishing "affinity groups" of administrators of the same rank but of different line offices within his organization. He asked people to be clear about their roles, using such role definitions as coach, honest broker, convener, recorder, and reporter. He expected "no tourists, and no prisoners" at meetings, to the discomfort of some observers, who thought their exclusion a sign of rudeness. He had used identical language and techniques as technical director at the Naval Ordnance Station at Indian Head and could point to successes there in building a more mission-oriented science and technology facility.[32]

To make the new culture explicit at the Office of New Production Reactors, he selected four paragraph-length passages from the works of Admiral Rickover that discussed shared basic principles. These remarks on technical competence, unrelenting dedication, individual responsibility, and intellectual honesty—all drawn from various statements Rickover made before Congress—were printed as mottoes and distributed to all ONPR employees. Many posted the quotations in their offices. In this rather specific fashion, Monetta graphically linked himself with the Rickover tradition and established that within this office of DOE, a cultural change was well under way. The specific management culture in ONPR began to emerge around a unique set of rules and practices. Although echoing some of Watkins' concerns, it went in a separate direction, defined by its own director and shaped by its rapid growth.[33]

Expert Choices without Special Pleading

The tasks confronting ONPR were straightforward but large in scale. First, a "down-select," or choice, had to be made among the various design firms and architectural and engineering (A&E) contractors hoping to work on each of three designs—heavy-water, high-temperature gas-cooled, and the

WNP-conversion light-water. DOE had made no commitment to complete the WNP, but ONPR investigated the design of an appropriate lithium deuteride target element that could be used in WNP light-water reactors to provide for a third technological approach, along with the heavy-water and high-temperature gas-cooled approaches. Although conversion of partially completed WNP light-water reactors was a less expensive path than construction of either a complete heavy-water or gas-cooled reactor, research and development of target elements had to precede a determination as to whether the path was viable.[34]

ONPR had to prepare the documentation necessary for a massive environmental impact statement for each of the three technologies for each of the three sites—in effect, for nine different possibilities.[35] Multiple hearings on the impact of the reactors upon the regions near each of the three sites had to be held.[36] Internal "requirements documents" outlining the specifics to be covered in conceptual design work were developed following a pattern of preset criteria much like that begun with N reactor at Hanford, but now much more elaborate and thorough. The selected design firms developed preliminary conceptual design studies on the various systems in each reactor type, and ONPR evaluated the studies in detail. Analysis incorporated probabilistic risk assessment of subsystems to determine overall risk.

Monetta and his team sought to provide the information necessary to select the best site and the best technology on grounds that were free of political pleading. ONPR established separate divisions within the office for each of the three technologies. A natural and fostered internal competition between the three approaches flourished, embodied in the three divisions. A fourth office dealt with safety and quality assurance. Each division director was assisted by a technical director; each worked with a cluster of contractors and support groups drawn both from outside contractors and from specialists within the DOE weapons complex and laboratories. Outside senior consultants provided prestigious and well-informed judgments as well as formalized links to prior generations of nuclear engineers, physicists, probabilistic risk specialists, and nuclear facility managers. Some of them had served on the distinguished NAS-NAE panel convened to study DOE's reactors after Chernobyl.[37] ONPR program management offices were established at each of the three sites, and a telecommunications net operated for rapid exchange of information between headquarters and the sites.

In October 1989 DOE entered into negotiations with two corporate teams for design of the heavy-water reactor (HWR) and a third team for the modular high-temperature gas-cooled reactor (MHTGR). The third option, the light-water reactor, did not require a full-blown A&E team but only contracted studies of the target design.

In order to decide on design and A&E firms, ONPR held off-site meetings at the well-equipped School of Seamanship at Piney Point, Maryland, which provided meeting rooms, dormitory accommodations, dining quarters, and, above all, isolation. There, the ONPR teams worked long hours to choose design and A&E firms. The teams reduced the design contractor groups to two: EBASCO, a consortium working on the heavy-water design, and CEGA, the consortium of Combustion Engineering and General Atomics that worked on the high-temperature gas-cooled reactor. As A&E firms, the group selected Bechtel for the heavy-water reactor model and Fluor Daniel for the high-temperature gas-cooled model. Further studies continued on the types of lithium deuteride ceramic-metallic, or "cermet," targets that could be used in light-water reactors.

Through 1990 and 1991 ONPR worked closely with the contractors, developing collections of materials for the site and technology selection processes and holding extensive public reviews in South Carolina, Idaho, and Washington as part of a NEPA-driven EIS process. Fully aware of the support and opposition that production reactor planning had inspired in the early 1980s, Monetta was determined to make the technical choice through a method that was legally unassailable.

The emergence of a team approach was sometimes made difficult because the internal competition between the conceptual designs reflected the more heated external alignments of corporate and regional technopolitical factions. The specialists in the MHTGR group thought their technology superior to the HWR approach, which they regarded as an outmoded design from the 1950s, and both thought the LWR approach held no promise of real progress in reactor design.

The corporations advocating the two leading designs each regarded their own approach as technically superior and the other as backed only by those who stood to gain from it professionally or financially. General Atomic's HTGR proponents argued that their design was not only inherently safe but that it offered excellent prospects as a model for a new generation of safe reactors for electric power generation.[38] Not to be outdone, EBASCO's vice president for technology, Robert Iotti, claimed that the heavy-water model "will be the safest reactor ever built," that operators could walk away in case of an accident while automatic features closed down the reactor, and that gas-cooled reactors had not been efficient.[39] Inside ONPR, sentiments were less hostile and more restrained but still competitive.

Other issues also generated internal technopolitical debate. Some of the safety specialists remained skeptical of oversimplistic use of PRA figures, while others believed that PRA was an excellent and necessary tool for making design decisions. PRA methods, they argued, should not only be used in evaluating subsystems but should be incorporated in plans for reliability, availability, maintainability, and inspectability (RAMI). Prior use of PRA in the design work for commercial reactors was closely studied. The method was used, without any exaggerated claims for infallibility, to help sort through design alternatives.[40]

Monetta remained above the fray, relying on the affinity groups and the overall ONPR team effort to harness the individual competitive energies and direct each towards the central program mission. As a means of reducing the naturally emerging loyalty of personnel to one design or the other, Monetta and his management team assured the staffs of the various internal design groups that once a conceptual design had been chosen, there would be guaranteed employment for all ONPR staff as the office switched from selection to operation of engineering and construction contracts. Those who had worked in the offices concerned with the eliminated conceptual designs could anticipate transfer to new positions inside other parts of the office when the reactors were to be built. In this way, personal careers would remain linked to the success of the total NPR program rather than to the success of a particular conceptual design. Even so, it was only natural for ONPR staff to favor the design on which they were working.

Conceptual designs, safety studies, technical issues, financial considerations, RAMI plans, and EIS work all were collected for the decision process. Out of the research and the submissions by contractors, ONPR generated documentation and compressed it into comprehensive secretarial briefing books. ONPR teams then provided the findings to Watkins during fifteen presentations. The plan was that he would consider the information, make his decision on the basis of technical merit by December of 1991, and then present it to the president.

Although the congressional delegations from the losing site or sites could be expected to complain, they would not have a legal or technical basis for their complaints if the procedure worked as planned. Although the process was stopped short of a final decision, Watkins' staff did not fully recognize or acknowledge that if a final choice had been made and funds had been appropriated, a bitter and hard-fought round of close examination of the decision process and the political factors involving advocates and opponents would ensue. At that time, any appearance of favoritism would have been scrutinized; any allegation of impropriety would have received full airing. Several potential questions regarding the objectivity of Watkins' personal staff imperiled the attempt to carefully structure an objective process. Rumor and innuendo about the personal relationship between Watkins' assistant and an EBASCO lobbyist led her to publicly withdraw from the selection process after she had raised concerns that the evaluation be fair to the heavy-water design.[41]

While the objective and technical studies went forward inside ONPR, political jockeying among the representatives of the three sites continued, a clear indication that once the secretarial and presidential decision was announced, the losers would come forward to argue their case as forcefully as possible in other forums, including the press and Congress. The open struggle for the lucrative and prestigious task had only been postponed and muted, not eliminated, as evidenced by continuing, if short-lived, public controversies over procedural issues. Environmentalists criticized the House Armed Services Committee, whose subcommittee on nuclear weapons was chaired by John M. Spratt of South Carolina, for trying to short cut the technically objective process early in 1991. That committee added to the 1992 defense budget a "sense of the Congress" resolution declaring that South Carolina would be the best site for the new reactor and requiring DOE to freeze NPR funding for 90 days if Idaho were chosen, while explaining to Congress its choice. "You're getting the political decision before the scientific one," claimed Brian Costner of a South Carolina energy monitoring group. "It usurps the decision making process established by DOE. . . . It's the worst kind of policy making. There's no excuse for it." Costner called the action "a classic case of pork barrel."[42] In Idaho, advocates of that site saw the Spratt gambit as an attempt at a political "pre-emptive strike" and hoped it would backfire against South Carolina.[43]

At the same time that ONPR worked on developing information for Watkins to use in deciding on a site and a technology for the new reactor, Savannah River proceeded with plans to refurbish K reactor for a restart. Watkins first announced plans to bring K up to potential restart so that a tritium production capacity would be available and then to place the reactor on "warm standby" for future use. Some members of Congress believed the whole K restart effort was wasteful. If K were successfully operated, the need for a new production reactor would diminish, and it would be more difficult to argue for and obtain the multibillion-dollar funding required to design and build the next reactor. Despite such objections, Watkins continued with plans to restart K reactor. Before the facility could be restarted, however, massive modifications and repairs were required. A persistent problem remained with leaks in one of the 12 large heat exchangers in which the heavy-water-moderator coolant was cooled with light-water in a secondary loop. When restart was finally attempted late in 1991, 150 gallons of radioactive tritium-contaminated water flowed out through effluent to the Savannah River. Part of the expense involved in the K restart derived from an upgraded cooling system featuring a cooling tower to mitigate thermal pollution of the ponds and streams, together with a host of other technical improvements; in all, the expenditures on refurbishing K reactor were in excess of $900 million.[44]

Meanwhile, Monetta and his senior management group worked towards a specific deadline based on helping Watkins to reach a technically objective choice between the NPR options in December 1991 and presenting to the president the secretary's preference as to site and technology. A Record of Decision (ROD) was planned for announcement on Sunday, 29 December.

However, international events overtook the new production reactor.

The End of the Cold War and Decline of the Arms Race

Leonid Brezhnev served as party secretary and as successor to the power of Lenin, Stalin, and Khrushchev in the Soviet Union from 1964 to his death in 1982. In the last 5 years of his tenure, a few signs of an impending crisis, not taken too seriously by observers, emerged in the Soviet Union and its satellites. Increasing economic stagnation and corruption, discontent among troops bogged down in Afghanistan since 1979, and the emergence of Solidarity as effective peaceful opposition to the Communist regime in Poland gave lie to the official portrait of the triumph of socialism. When Mikhail Gorbachev was elected general secretary of the Communist party in March 1985, he was already identified as a representative of a younger generation of bureaucrats.

In 1989, the year in which President George Bush appointed Admiral Watkins and Watkins in turn appointed Monetta to direct the NPR program, there were further fundamental changes in the Eastern bloc. That year saw a multiparty election in the Soviet Union in which the Communist party-state apparatus suffered a serious defeat, the mass migration of East Germans via Hungary to West Germany, the fall of the Berlin Wall, and political changes that swept Communist regimes from most of the Eastern European satellite states. By the end of 1989 the Soviet Union had accepted the concept of reunification of Germany; the treaty achieving the unification was implemented on 1 January 1991.

In the midst of these changes, Gorbachev continued arms negotiations with the United States, signing a Strategic Arms Reduction Treaty (START I) in July 1991. Under the terms of the treaty, both the Soviet Union and the United States would reduce not only the delivery systems that had been addressed under the SALT II treaty but the total number of thermonuclear warheads as well. Conservatives in the Soviet Union, those opposed to the loss of power and to the sweeping economic and constitutional changes, tried to restrain Gorbachev early in 1991. In August of that year they mounted an abortive coup, holding Gorbachev under house arrest for a few days, just as he was about to sign a new treaty restructuring the Soviet Union. Between August and December 1991, Boris Yeltsin, president of Russia, emerged as the effective leader. The Soviet Union was replaced with a loose Commonwealth of Independent States of 11 of the 15 former Soviet republics.

This rapid change, first to a reforming regime, then to a completely different national and international structure, caught American policymakers by surprise. Although some commentators had predicted for decades that internal difficulties in the Soviet Union would bring about change, even as it was beginning almost none anticipated that it would end with the collapse of the whole Soviet regime.[45]

Quite suddenly, many of the basic premises of American foreign and defense policy became less relevant. For the nuclear weapons complex, the restructuring of the world had profound effects. Troop reductions in Eastern Europe and the political changes there in 1989 reduced the threat. When treaties were signed, beginning with the July 1991 START I agreement, reduction in total weapons had already begun.[46]

As the dramatic changes unfolded, they required repeated rethinking of the American defense budget and the nuclear weapons complex. As the Cold War seemed to decline, disarmament advocates urged dropping the production reactor effort, but DOE continued to hold to its schedule through the summer and early fall of 1991.[47] A few editorialists, like one near Savannah River in Aiken, South Carolina, argued that the increasing instability of the

Soviet Union required even greater vigilance and nuclear preparedness, calling the state governor's opposition to a new production reactor on environmental grounds evidence that he was a "wimp of the first order."[48]

Increasingly, the most serious issue in the weapons complex became the effort to manage the cleanup of polluted and radioactive facilities, rather than replacing the badly weakened tritium-producing capacity.[49] With the reduced need for nuclear weaponry, the urgency to reach a Record of Decision regarding new production reactors declined sharply in 1991, just as ONPR was making its pre-ROD presentations to Watkins.

New Production Reactors Canceled

On 1 November 1991, Watkins suddenly and unilaterally announced that the scheduled date for the Record of Decision regarding production reactor technology and site selection, 29 December 1991, was to be set back by 2 years, until the end of 1993. Surprised at the unexpected news of the change in schedule, Monetta submitted his resignation, as did John C. Tuck, undersecretary of energy. Both were shocked at the abrupt decision and frustrated at not being informed prior to the public announcement. Their joint resignations caused a flurry of press attention.[50]

Sen. Sam Nunn, chair of the Senate Armed Services Committee, was also stunned. The Senate had urged DOE to move quickly on NPR planning and had opposed Watkins' work on restarting K reactor at Savannah River. If Watkins' decision to put off the NPR decision resulted from contractor lobbying of Congress, Nunn pointed out, that decision was unjust. Those contractors, Nunn stated, "like any one else, have an absolute First Amendment right to petition Congress and to express their views. Our national policy depends on the input from a wide variety of sources, not just the Secretary of Energy. The Constitution vests these responsibilities in more places than your office."[51]

Watkins assigned the work of the director of ONPR as well as that of acting undersecretary to Thomas Hendrickson, a nuclear engineer who had formerly served under Rickover in the nuclear Navy and who had worked with the nuclear firm Burns and Roe.[52] Hendrickson, who shared a Rickover-inspired dedication to technical excellence from experience in the Navy, was skeptical of formal management methods. He frankly relied on his knowledge of individuals in the nuclear group within DOE and a commonsense approach to budget and administrative matters rather than on a more structured theory of managerial science. Like Monetta, he held an undergraduate degree in a technical field and a graduate degree in administration.

In early 1992, as he began management of ONPR, Hendrickson accepted the concepts of lower urgency and the postponed decisions that Watkins had decided upon. Hendrickson anticipated that if the NPR project survived through all the necessary oversight systems, a reactor could be brought on line in the year 2005. Rather than viewing the 13-year period as an indication of bureaucratic or political delay, he viewed such oversight as appropriate to prevent wasted funding. He recognized that changes in the nation's weapons configuration or stockpile requirements could reduce the urgency even further and that the schedule might readily slip again. In fact, the original schedule was abandoned.[53]

In 1992 Hendrickson undertook an organized and scheduled dismantling of the NPR effort. Over the next several months, DOE closed down the outstanding contracts with various firms that had been developing designs of the technologies and providing A&E work to the NPR effort. The department closed out or transferred to other internal units the last of the ONPR contracts early in 1993.[54] Hendrickson converted his role to that of administrator-caretaker as Watkins ordered the project to wind down. The personnel within ONPR shifted their careers, many taking positions elsewhere in the department, some moving to other agencies or out of government service, and some taking early retirement. Monetta moved on to a position in the Pentagon in the Office of the Secretary of Defense and later established Resource Alternatives, Inc. a consulting firm in Washington, D.C. Former Undersecretary Tuck joined the Washington law firm of Howard Baker.[55] The Office of New Production Reactors was disestablished, its records were archived, and the final "down-select," or decision as to preferred technology, was never announced.

With the elimination of MIRVs in the START II treaty, signed in June 1992, and with that treaty's sharp cuts in the total number of weapons, supplies of all strategic nuclear materials were more than adequate for the planned requirements. The cannibalization, or the "mining," of tritium from old weapons provided a supply of tritium to maintain the readiness of remaining weapons.[56] Plutonium, with its very long half-life, would never have to be produced to supply weapons. With the cannibalization process and arms reductions, a tritium shortfall would not occur until well into the twenty-first century.

Early in 1992, as Watkins announced plans to begin closing various parts of the weapons complex in response to the ending of the Cold War, he admitted that nearby communities would suffer from job loss, but emphasized that closings were necessary. "Let's declare victory and phase ourselves down responsibly," he said.[57]

Conclusion:
Supplying the Cold War Arsenal

The three production reactors at Hanford which produced the awe-inspiring weapons that abruptly ended World War II were the first of an eventual 14 production reactors in the United States. Two more reactors were added in the early years of the Cold War, as tensions mounted between the Soviet Union and the United States, bringing the total to five. After the Soviets exploded their first nuclear bomb, the United States decided to work towards a fusion weapon and undertook to build three more reactors at Hanford and five at Savannah River, capable of making tritium as well as plutonium.

The Manhattan Engineer District and its successor agencies were created out of existing communities of scientists, corporate executives, and army officers. The tensions between the free-wheeling academic style of the scientists and the security-conscious military men received much public attention in biographies and memoirs. But to harness the business community, further compromises had been required. The result was the evolution of Army Corps contracts into the structured government-owned contractor-operated facility, the GOCO. Only through such a contract vehicle could the government attract and hold with the necessary financial incentives long-term assemblies of scientists, technicians, administrators, and support staff. As policy changed over the following decades, those GOCO communities developed articulate political clout. Community leaders and spokesmen like Glenn Lee at Richland, Washington, soon spoke out for the contractors' employees. Congressmen and senators grew increasingly sensitive to the lobbying and voting power of the constituents clustered in Washington, Idaho, and South Carolina.

Decisions regarding plutonium and tritium supply could not simply be made behind closed doors. Corporate and public interest groups and a powerful group of senators and members of Congress watched nuclear policy closely, even if the more general public remained relatively unconcerned.

Meanwhile, the profession of nuclear engineering grew and evolved, reflecting the increased influence of those who first sought to turn reactors to ship propulsion and then to civilian power production. Experiments and demonstrations through the 1950s generated a variety of reactor cousins, designed to help supply the nation's and the world's need for plentiful electrical energy. In this context, the Atomic Energy Commission added its fourteenth production reactor, the odd hybrid, N reactor, which was a cross between a production reactor and a power reactor, optimized between two functions. No sooner was N reactor on line, however, than the United States achieved a sufficient stockpile of plutonium; thus, all but the four newest reactors were closed during a period of plutonium oversupply and political détente. As the older reactors closed, the new emphasis on peaceful uses of the atom pushed the total megawattage of power reactors higher than the total megawattage of the remaining small family of production reactors. For nearly a decade, through the 1970s, the four remaining production reactors supplied the nation's need for strategic materials.

The rise in tensions between the United States and the USSR following the Soviet invasion of Afghanistan and the strong defensive posture of the United States in the 1980s required advance planning to meet the forthcoming tritium shortfall and the increased need for plutonium in MIRV weapons. That anticipated shortfall became more imminent as the last of the 14 reactors closed forever. Increasing public distress at the environmental risk of both power reactors and production reactors accelerated those closures. Although some journalists and political critics believed as early as 1988 that the natural disarmament brought by tritium decay should be allowed to proceed in the United States whether or not the Soviets agreed to halt tritium production, the mission to build a reactor was not abandoned. Secretary of Energy Herrington left the task to his successor, Admiral Watkins; as head of the newly created Office of New Production Reactors, Monetta moved decisively to get a single design that he could present as the best possible one, uninfluenced by political pressures or special interests.

Despite these efforts to reach an unbiased decision, technopolitics continued unabated, as the backers of the two leading conceptual designs focused their efforts on making good presentations and developing arguments useful against the opposition, and as congressional representatives of the potential sites continued to maneuver for position. Had a decision been announced, the corporate and political backers of the excluded designs would have mounted a vigorous public relations campaign to reconsider in congressional and media forums the relative virtues of the systems. To assume otherwise would be naive.[1] Yet such a discussion could have taken place against a background of objectively measurable financial, engineering, and scientific data collected in a fair fashion.

Planning for new production reactors moved from a squabble over patronage into a managed decision environment that demonstrated how data for a difficult technical choice which could generate billions of dollars of employment might be gathered and developed both objectively and rapidly. Systems analysis could be funded on three separate conceptual designs, the merits of the developments could be measured and reviewed by experienced and independent experts, and the final executive choice could be based on financial and design criteria, which although presented by advocates, had been collected without favoritism. Whether the ultimate decision could be equally nonpolitical was doubtful, and the question was never put to the test.

With the end of the Cold War, there was no longer a pressing need for an assured tritium production capacity. The effect of the changed international situation was to move new production reactors for strategic nuclear materials to a much lower priority. The technology, the capacity, and the planning, while necessary to the maintenance of a deterrent nuclear arsenal at the height of the Cold War from the 1950s through the early 1980s, quite suddenly came to an end.

Two Department of Defense consultants had suggested in 1990 that "virtual deployment"—that is, active planning towards a future weapons development—could influence the behavior of adversaries. In retrospect, the ONPR effort seems to have been just such a program. The Office of New Production Reactors, with its high-profile, increasing design expenses, and its widespread public hearings over environmental impact, like the Strategic Defense Initiative of the Reagan administration, may have served a purpose in international negotiations even though no foundation was dug, no concrete poured. Together with the temporary but well-publicized effort to restart K reactor at Savannah River, ONPR demonstrated that America could keep up its nuclear deterrent if needed. The Soviet regime, in its last year of power, had ample evidence that if no START agreement were reached, the United States was capable of devoting billions of dollars to tritium production and to a continued arms race.[2]

Although the next generation of production reactors was suddenly aborted, the project left several legacies. One legacy was a method which might be employed in other competitions for such massive projects. Multibillion-dollar engineering feats of the future in which more than one design might be appropriate required a procedure that allowed for selection on grounds of technical merit, rather than on the basis of political influence, and that at the same time permitted the general public and interested parties to participate through the open methods developed since the days of World War II. The structure and procedures of ONPR could serve as a model for efforts to control or at least mitigate the effect of patronage battles in favor of efficient technological choice for projects of such magnitude. Yet, as always, those technological choices would be subject to the pressures of professional identification, theoretical and engineering styles, corporate cultures and corporate interests, as well as to the power of interested members of Congress. Engineering, budget analysis, and management could only go so far as tools to reduce the difficulties of technopolitical choice.

Dealing with the legacy of generations of radioactive waste and hazardous effluent, the vast establishments at Savannah River and Hanford converted their priorities to environmental cleanup and restoration. The effort promised to cost far more than the original cost of constructing the weapons complex. For nearby residents and downwinders, the nuclear heritage remained a public health concern.

Still another legacy of the nuclear arsenal and the production reactors that had fueled it was the technology itself, readily imitated by nations seeking new weapons with which to exert their power. At the heart of the nuclear arsenal were the reactors that produced the strategic materials, and reactor design and technology had spread around the planet. Representatives of the international community, working through the International Atomic Energy Agency, sought to determine the possible date at which Iraq and North Korea would join the world's nuclear powers, and the nuclear threshold status of those two states remained critical international issues through the early 1990s. Power-generating and research reactors in both Iraq and North Korea, as well as in more politically stable nations, could be readily diverted from their ostensible peaceful purposes to play the role of production reactors.

If and when the United States decided to resume tritium production to resupply its nuclear arsenal, the massive legacy of technical plans and designs created over the 1980s and 1990s was available.

APPENDIX
Production Reactor Families

Since all of the U.S. production reactors were designated by letters instead of names, the reader may find a condensed presentation of data on the reactors useful for reference to place particular reactors in context. Table A.1 shows the groups of reactors. Brief technical sketches of each group follow. A list of auxiliary reactors is given in Table A.2.

Hanford Reactors

The Wartime Round
Du Pont completed B, D, and F reactors between September 1944 and February 1945. The three reactors were designed at 250 MW and upgraded in the postwar years to over 435 MW by 1951. By 1963, B was rated at 1,940 MW, D at 2,005 MW, and F at 1,935 MW. Fuel slugs were 1.45 inches in diameter and 8.5 inches in length. The reactors were cooled by single-pass river water and moderated with graphite. The biological shielding consisted of laminated masonite and steel; the thermal shielding was concrete. (The basics of this design were applied for five more reactors built between 1948 and 1955 at Hanford, described below.) B reactor went into operation on 26 September 1944; D reactor on 17 December 1944, and F reactor on 25 February 1945. In order to preserve one reactor as an emergency backup during the postwar crisis over graphite expansion, B reactor was closed in 1946 and reopened in 1948. F reactor was shut down on 25 June 1965; D on 26 June 1967, and B on 12 February 1968. B reactor was designated a Historic Mechanical Engineering Landmark in 1976.

The Postwar Round
General Electric followed the designs of B, D, and F reactors for construction of H and DR. H opened 10 October 1949, after only 18 months of construction. It was upgraded to 500 MW by 1951 and to 1,955 MW by 1963. DR was built as a replacement for D reactor and was designed originally to take over the waterworks of D. Later, a separate waterworks was built for DR, and operation of DR had to wait for completion of this facility. Originally designed at 250 MW, DR went into operation on 3 October 1950; it was upgraded to 500 MW by 1951, and to 1,925 MW by 1963. DR was shut down on 30 December 1964; H was shut down on 21 April 1965.

Table A1. U.S. production reactor families

Hanford reactors	Savannah River reactors
Wartime round	
1944: B	
1944: D	
1945: F	
Postwar round	
1949: H	
1950: DR	
Korean War round	*"Little Joe" response round*
1951: C	1953: R
1955: KW	1954: P
1955: KE	
	Savannah second round
	1954: L
	1954: K
	1955: C
Dual-purpose or hybrid	
1964: N	

Note: Dates are year of reactor startup.

The First Korean War Round

General Electric built C reactor rapidly, following early designs, but on a larger scale, to meet the emergency of the Korean War. It went into operation on 18 November 1952. Originally designed at 750 MW, it was upgraded to 2,310 MW by 1963. C was shut down in 1969.

KE and KW were "Jumbo" reactors, designed at 1,800 MW(t) when they opened in early 1955. Both were upgraded to 4,400 MW by 1963. Both had systems of space-heating, using an ethylene-glycol heat exchange system to reduce utility costs in heating the reactor work areas. Because of faulty prestart inspection, KW suffered a process tube water leak and overheating 17 hours into its first operation in January 1955, requiring extensive repairs before restart later that year. KW was shut down in 1970, and KE in 1971. In the 1990s, the slug-storage tanks of water at KW were used for storage of unprocessed fuel slugs from other reactors.

The Dual-Purpose Reactor

N reactor at Hanford was quite different from her older sisters. Dubbed "N" as an abbreviation for "new production reactor," N was designed, beginning in 1958, by General Electric as a dual-purpose, or convertible, reactor, to produce both plutonium and heat for steam turbines and electricity. N reactor was graphite-moderated and cooled by pressurized water at about 250°F.

Table A.2. Auxiliary test and experimental reactors at Hanford and Savannah River

Site/reactor	Years of operation	Power level
Hanford		
Hanford 305 Test Reactor	1944–1972	Unknown
Plutonium Recycle Test Reactor	1960–1969	Unknown
Neutron Radiograph Facility	1977–	250 kW
Fast Flux Test Facility	1980–	400 MW
Savannah River		
Savannah River Test Pile ("305")	1953–1988	50 MW
Heavy-Water Components Test Reactor	1962–1964	61 MW
Lattice Test Reactor	1967–1979	0.5 kW
Process Development Pile	1953–1979	0.5 kW
Standard Pile	1953–1979	To 10 kW

At 2.4 inches in diameter by 26 inches long, the fuel elements in N reactor were larger than those of earlier reactors. They were coated in 0.03 to 0.04-inch-thick zircalloy (ZR-II) cladding. The design rating was 3,950 MW(t), or 863 MW(e).

The reactor began operations in 1964 and produced steam for nearby power plant owned and financed by Washington Public Power Supply Service, a consortium owned by rural electric cooperatives and municipal power companies. It was placed in standdown on 12 December 1986, partly in response to concerns over the fact that it was the last large graphite-moderated reactor in the United States when the graphite-moderated Chernobyl reactor disaster occurred 26 April 1986. N reactor was never reopened, being placed in "cold standby" on 16 February 1988.

Savannah River Reactors

"Little Joe" Round

In response to perceived increased needs for tritium following a presidential decision to design a fusion weapon after the Soviets detonated their first atomic weapon, dubbed "Little Joe," in August 1949, two heavy-water-moderated reactors, designated R and P, were planned and constructed by Du Pont at Savannah River. R began operation on 28 December 1953, and P reactor was started on 20 February 1954. The reactors consisted of large steel vessels in which the fuel and target elements were inserted vertically. The moderator was heavy water or deuterium (^2H_2O), which also served as the primary coolant. The deuterium was cooled in a heat exchanger by ordinary light water, itself released to cooling ponds. Rated at 383 MW at their opening, the reactors were raised to a nominal level of 2,000 MW by 1963, operating in the range of 2,300 MW to 2,600 MW. The fuel elements were normally clad in 0.03-inch aluminium and were 4 inches in diameter. A great variety of experimental fuel and target elements were later designed to fit in the tubes, along with special target elements for the production of isotopes. (These same parameters were applied to the second round of reactors at Savannah River.) R reactor was closed on 15 June 1964 and cannibalized for parts; P reactor was closed on 17 August 1988.

Savannah Second Round

Three more production reactors were built at Savannah over the period 1951–55, following the designs of R and P. Like the first two, the later three heavy-water-moderated and-cooled reactors were originally rated at 383 MW, and upgraded to a nominal 2,000-MW level. L and K reactor began operations in July and October 1954, respectively; C reactor began operations March 1955. L was placed on standby on 18 February 1968 and after lengthy hearings and refurbishing, it was restarted in October 1985.

In March 1987, P, K, L, and C reactors were placed on 50% power, and by the end of 1988, all were closed. Plans for a restart of K reactor involved redesign of the primary coolant system, construction of a cooling tower for the water from the heat exchangers, and extensive retraining of personnel and other modifications. After a brief demonstration operation in 1993, K reactor was placed on permanant shutdown status.

Auxiliary Reactors

At both Hanford and Savannah River, a number of auxiliary test and experimental reactors were built over the years. A partial listing is given in Table A.2.

NOTES

The following abbreviations are used in the notes. Other, more general abbreviations can be found in the list of acronyms and abbreviations in the front matter.

Greenewalt Notes	Crawford H. Greenewalt's Notes, 1942, NARA, RG 326, History Division, DOE, box 2, folder 6
Hagley	Hagley Accession 1957, Hagley Archives, Wilmington, Delaware
Hanford RHA	Hanford (Wash.) Records Holding Area
NARA	National Archives and Records Administration, Washington, D.C.
RG	Record Group
AEC 1140	AEC 1140, "History of Expansion of AEC Production Facilities," DOE Archives, RG 326, Secretariat, box 1435, folder I&P 14, History
MH&S 16-4	Hanford RHA, box 15200, folder Medicine, Health and Safety, 16-4
AEC 24/22	AEC 24/22, Director of Production, "Characteristics of 'X' Reactors," 19 August 1952, DOE Archives, RG 326, Secretariat, Box 1282, Folder PLBL, Hanford
NPR Status Information	NPR Status Information, copy packet in EG&G Collection, DOE Archives (not archived at time of use)

Chapter One: Inventing Atomic Piles

1. For a popular and personal essay regarding scientific and engineering accomplishments and their bearing on weapons and strategy, see Vannevar Bush, *Modern Arms and Free Men: A Discussion of the Role of Science in Preserving Democracy* (New York: Simon and Schuster, 1949); for a modern comparative analysis of war production, and statistics on aircraft production, Alan S. Millward, *War, Economy, and Society, 1939–1945* (Berkeley and Los Angeles: University of California Press, 1977).

Two specialized works on technical and industrial achievements are R. H. Connery, *The Navy and Industrial Mobilization in World War II* (Princeton: Princeton University Press, 1951), and Civilian Production Administration, *Industrial Mobilization for War: History of the War Production Board and Predecessor Agencies, 1940–1945* (Washington, D.C.: GPO, 1947).

2. The literature on organizational and corporate culture has proliferated since the early 1980s. This work draws on the approach suggested by Louis Galambos, in the seminal article "The Emerging Organizational Synthesis in Modern American History," *Business History Review* 44 (Autumn 1970): 279–90.

3. Brian Balogh, *Chain Reaction: Expert Debate and Public Participation in American Commercial Nuclear Power, 1945–1975* (New York: Columbia University Press, 1991), 1–20.

4. Vannevar Bush, *Pieces of the Action* (New York: William Morrow, 1970), 53, 55. Here and elsewhere, Bush commented on the military view of scientists and engineers, as well as on the relationships between science and engineering. For a review of the evolution of the successor agencies, see Alice Buck, *A History of the Atomic Energy Commission* (Washington, D.C.: DOE, July 1983), 8.

5. For the Navy's Powder Factory at Indian Head, Maryland, see Rodney Carlisle, *Powder and Propellants: Energetic Materials at Indian Head, Maryland, 1890–1990* (Washington, D.C.: GPO, 1991).

6. The international fraternity is discussed in Richard Rhodes, *The Making of the Atomic Bomb* (New York: Simon and Schuster, 1986), 104–33.

7. Speaking of reactor engineers, Richard Rhodes noted that "Fermi was only then inventing that specialty." Rhodes, *The Making of the Atomic Bomb*, 432.

8. Leslie R. Groves, *Now It Can Be Told: The Story of the Manhattan Project* (New York: Harper, 1962), 44.

9. See David A. Hounshell, "Du Pont and the Management of Large-Scale Research and Development," in Peter Galison and Bruce Hevly, eds., *Big Science: The Growth of Large-Scale Research* (Stanford, Calif.: Stanford University Press, 1992), 236–61.

10. J. L. Heilbron and Robert W. Seidel, *Lawrence and His Laboratory: A History of the Lawrence Berkeley Lab* (Berkeley and Los Angeles: University of California Press), 1:240. The young scientists remarked at the unfamiliar, industrial nature of their work. Albert Wattenberg compared his appearance after a day on the graphite construction work both to a coal miner and to a minstrel actor. Clearly, the work was a cultural shock. Corbin Allardice and Edward R. Trapnell, *The First Pile*, Report No. TID 292 (Oak Ridge, Tenn.: AEC, 1949), 4.

11. A. H. Compton to F. Jewett, 17 May 1941, National Archives and Records Administration (NARA), RG 227-OSRD, S-1 Materials, box 6, Files of Lyman J. Briggs, folder National Academy of Sciences; Allardice and Trapnell, *The First Pile*, 3. Rhodes unravels the intricacies of developments through 1941 in much greater detail than is presented here. Rhodes, *The Making of the Atomic Bomb*, 365–88.

12. Leo Szilard-Enrico Fermi Correspondence, five letters, July 1939, and Leo Szilard, "Divergent Chain Reactions in Systems Composed of Uranium and Carbon," in *The Collected Works of Leo Szilard*, ed. Bernard T. Feld and Gertrude Weiss Szilard, 2 vols. (Cambridge: MIT Press, 1972), 1:193–95, 216–56.

13. Arthur Holly Compton, *Atomic Quest* (New York: Oxford University Press, 1956), 87–89.

14. Ibid., 89, 102.

15. Richard Hewlett and Oscar Anderson, *The New World, 1939–1946,* vol 1 of *A History of the Atomic Energy Commission* (Washington, D.C.: AEC, 1962), 174–75; see also 180 n. 9.

16. Allardice and Trapnell, *The First Pile,* 3; Peterson Report, 29 September 1942, organization table, 13–14, NARA, RG 77, entry 5, box 26, folder 600.12, Thermal Diffusion Project; on 29 September 1942, Oppenheimer was shown as working under Compton, heading up the Fast Neutron Reactions group that was dispersed over a number of universities.

17. Hewlett and Anderson, *The New World,* 178–79. The Engineering Council was renamed the Technical Council in July 1942.

18. Emilio Segré, *Enrico Fermi, Physicist* (Chicago: University of Chicago Press, 1970), 130; Hewlett and Anderson, *The New World,* 179–80.

19. Vincent Jones, *Manhattan: The Army and the Atomic Bomb* (Washington, D.C.: Center for Military History, U.S. Army, 1985), 19. The Corps called the work the Development of Substitute Materials (DSM) project. Marshall, operating out of New York City, then changed the code name to the Manhattan Engineer District. See Compton, *Atomic Quest,* 107; Rhodes, *The Making of the Atomic Bomb,* 426.

20. Rhodes, *The Making of the Atomic Bomb,* 423–27; Hewlett and Anderson, *The New World,* 104–8.

21. Rhodes, *The Making of the Atomic Bomb,* 412, 422–24; Compton, *Atomic Quest,* 105; Hewlett and Anderson, *The New World,* 107–8.

22. Hewlett and Anderson, *The New World,* 181.

23. David A. Hounshell and John Kenley Smith, *Science and Corporate Strategy: Du Pont R&D, 1902–1980* (New York: Cambridge University Press, 1988), 333, 339.

24. L. R. Groves to R. Williams, 12 April 1945; Williams to Groves, 3 May 1945, with enclosures. Both in Hagley Archives, Wilmington, Del., Hagley Accession 1957, box 1, folder 1. (Hereafter all citations to "Hagley" are to this facility and accession.) For Groves's substantiation of this information, see Groves, *Now It Can Be Told,* 56–57.

25. Groves, *Now It Can Be Told,* 57.

26. For "stepwise," see de Right to Williams, 24 April 1945, and chronological details from TNX History, 1–13, both in Hagley, box 1, folder 1. The finalized contract, W-7412, eng-1, signed by Col. K. D. Nichols for the Corps of Engineers and by E. B. Yancey, general manager of Du Pont's Explosives Department, was dated 22 November 1943, nearly a year after the decision to go ahead. Hanford Records Holding Area, Hanford, Washington (hereafter Hanford RHA), box A-365, folder Early Hanford History.

27. Hewlett and Anderson, *The New World,* 186–87.

28. TNX History, 28 December 1944, 22.

29. Groves, *Now It Can Be Told,* 58.

30. Williams to Groves, 3 May 1945, and contract W-7412, eng-1, 22 November 1943, both in Hanford RHA, box A-365. In fact, government auditors eventually disallowed $0.33 of the $1.00 fee on the grounds that the contract did not run for

the full period allowed; Groves noted that the executives accepted the lower fee with a sense of humor. Groves, *Now It Can Be Told*, 59.

31. Rhodes, *The Making of the Atomic Bomb*, 442. The heat from CP-1 simply radiated into the cold air of the unheated squash court. The first pile was never operated long enough for continued cooling to become an engineering concern, and there was no positive cooling system. In later years, after the proliferation of nuclear reactors led to a taxonomy based on function, moderator, and coolant, CP-1—the grandfather of them all—would have been described as an experimental, graphite-moderated, convection-air-cooled, or "no-coolant," reactor. An early postwar text listed the coolant as "none." Raymond Murray, *Introduction to Nuclear Engineering*, 2d ed. (Englewood Cliffs, N.J.: Prentice Hall, 1961 [1954]), 128–29.

32. Emilio Segré noted that Fermi had already achieved $k > 1$ in the exponential piles; thus, the historic December 2 experiment was more of a demonstration than a proof as far as Fermi was concerned. Segré, *Fermi*, 129. Of the reported 29 experimental small piles built prior to CP-1, the first value of $k > 1$ was recorded on experimental pile 9. In effect, one of the "subcritical assemblies" had already reported criticality. See Richard E. Nightingale, ed., *Nuclear Graphite* (New York: Academic Press, 1962), 1, and Rhodes, *The Making of the Atomic Bomb*, 435, 439.

33. Albert Wattenberg, "December 2, 1942: The Event and the People," in Robert G. Sachs, ed., *The Nuclear Chain Reaction—Forty Years Later* (Chicago: University of Chicago Press, 1984), 43–53; Samuel Allison, "Ten Years of the Atomic Age," *Bulletin of the Atomic Scientists* 9 (February 1953): 8–10.

34. Wattenberg, "December 2, 1942," 43–53. Samuel Allison remembered that Greenewalt's attendance at the squash court demonstration was a lucky accident in that "Greenewalt happened to be visiting with Compton and came away from CP-1 quite impressed." See Allison, "Ten Years of the Atomic Age," 9. In his notes, Greenewalt described the fortuitous chance of being present and called the experience "thrilling." C. H. Greenewalt's Notes, 1942, 3:58, (hereafter cited as "Greenewalt Notes"), NARA, RG 326, History Division, DOE, box 2, folder 6, job 1346.

35. This exchange has been quoted many times with various levels of accuracy. Compare Allardice and Trapnell, *The First Pile*, and Compton, *Atomic Quest*, 144. Allardice, probably working from oral legend in 1949, reported Compton's historic quip slightly more crisply than Compton remembered it in 1956. Allardice's version also appears in his "The First Pile," *Bulletin of Atomic Scientists* 18 (December 1962): 19–24.

36. Rhodes, *The Making of the Atomic Bomb*, 436.

37. For *slow-downer*, see Greenewalt Notes, 1:2. The original derivation of *scram* was mentioned in a conversation between Rodney Carlisle and Albert Wattenberg at the Conference of the American Physical Society, Washington, D.C., 23 April 1992.

38. Hewlett and Anderson, *The New World*, 193–94.

39. Greenewalt Notes, vol. 1, 16–17 December 1942.

40. Ibid., 28 December 1942.

41. Ibid., 5 January 1943.

42. Groves, *Now It Can Be Told*, 70–71.

43. Ibid., 73; Report of G. H. Giroux, Power Consultant, 28 November 1942, and E. G. Ackert to L. R. Groves, 5 January 1943, both in NARA, RG 77, entry 5, box

72, folder 600.03. Nine of the sites rejected from map study in the states of Washington, Idaho, and Oregon were Moses Lake, Odessa-Wall Lake, Coeur d'Alene, Priest River, Horse Heaven Hills, Des Chutes River, Bend, Umatilla, and southeastern Oregon. "Areas Studied from Maps but not Suitable for Project," Hanford RHA, box C-309, folder 314.7, Site Notes.

44. Groves, *Now It Can Be Told*, 73–74; E. I. du Pont, Engineering Division, "Special Plant Site Location Investigation," 2 January 1942 (copy 2 of 10), NARA, RG 77, entry 5, box 72, folder 600.03 (copy 9 of 10 is stored at the Hanford RHA, box C-309.) In a 1945 War Department press release describing the career of Colonel Matthias, the site selection was attributed to him with no mention of his Du Pont teammates. Press Release, War Department, Hanford RHA, box A-717, folder Col. Matthias Info.

45. G. H. Giroux to Col. F. T. Matthias, 6 January 1943, Hanford RHA, box C-309, folder 314.7, Site Notes. Giroux referred to the Hanford site as the White Bluffs area.

46. Groves, *Now It Can Be Told*, 74–75; Hanford Project, Supplemental Gross Appraisal, Gable Property, Grant, Franklin, and Benton Counties, State of Washington, 23 January 1943, NARA, RG 77, entry 5, MED Decimal Files, box 51, folder 319.1. The 9 February date for formal acquisition is from Hanford Fact Sheet, n.d., Public Affairs Office, Hanford Area Office, DOE.

Chapter Two: Building Reactors at Hanford

1. Greenewalt Notes, vol. 1, 28 December 1942.

2. Ibid.; See David A. Hounshell and John Kenley Smith, *Science and Corporate Strategy* (New York: Cambridge University Press, 1988), 339–40 on the continuing animosity of Wigner towards Greenewalt.

3. Ibid.

4. Ibid.; Richard Hewlett and Oscar Anderson, *The New World*, 1939–1946 (Washington, D.C.: AEC, 1962), 191.

5. Greenewalt Notes, vol. 1, 5 January 1943; Richard Rhodes, *The Making of the Atomic Bomb* (New York: Simon and Schuster, 1986), 486–87.

6. Greenewalt Notes, vol. 1, entries for 6, 8, 9, 12 January 1943.

7. Ibid., 12 January 1943.

8. Hewlett and Anderson, *The New World*, 194, 197.

9. Greenewalt Notes, vol. 1, 16 February 1943.

10. Williams to Groves, 3 May 1945, Hagley, box 1, folder 1; Greenewalt Notes, vol. 1, 16 January 1943, 31 December 1942.

11. Arthur V. Peterson to Groves, 23 August 1943, NARA, RG 77, entry 5, box 75, folder 600.12, Projects and Programs.

12. Compton to Whitaker, 7 December 1943, NARA, RG 77, entry 5, box 75, folder 600.12, Projects and Programs.

13. "Program of the Clinton Laboratories (as of Dec. 1, 1943)" (also dated 26 January 1944), NARA, RG 77, entry 5, box 75, folder 600.12, Projects and Programs.

14. Draft of Manhattan District History, 19 November 1945, Hanford RHA, bk. 4, vol. 2, pt. 2. Some of the operators who were present at the startup of the Han-

ford piles were graduates of the training phase at X-10. Conversation of Rodney Carlisle with Don Lewis, at B Reactor, Hanford, Washington, 2 June 1992.

15. Vincent Jones, *Manhattan: The Army and the Atomic Bomb* (Washington, D.C.: Center for Military History, U.S. Army, 1985), 210.

16. Greenewalt Notes, vol. 1, 22 January 1943.

17. Ibid., 23 March 1943.

18. Ibid., 20 May 1943. Concern for the fish is expressed in a memo to the file, 18 September 1945, Hagley, box 58, bk. 14, Operations of Hanford Works, Technical Department, pt. 1.

19. Greenewalt Notes, vol. 1, 28 May 1943.

20. Rentenbach to Col. Franklin T. Matthias, 24 January 1946, Hanford RHA, Manhattan District History (Smyth Draft), bk. 4, Pile Project, vol. 5, Construction, box C-309, 9.15–9.16.

21. The decision to build at least one pile without refrigeration apparently sprang from a concern to get one up and running as soon as possible, during the winter months when Columbia River water was naturally cold. Greenewalt Notes, vol. 1, 11 May 1943.

22. Jones, *Manhattan*, 215–16.

23. Greenewalt Notes, vol. 1, 24 June 1943.

24. Compton to Groves, 5 February 1943, NARA, RG 77, entry 5, box 76, folder 600.12; Richard E. Nightingale, ed., *Nuclear Graphite* (New York: Academic Press, 1962), 4; Walter C. Patterson, *Nuclear Power* (Hammondsworth, Eng.: Penguin, 1976), 43.

25. Compton to Wigner, 23 July 1943, NARA, RG 77, entry 5, box 59, folder 333.5, Investigations.

26. Peterson to Groves, 13 August 1943, NARA, RG 77, entry 5, MED Decimal Files, box 59, folder 333.5, Investigations.

27. Greenewalt Notes, vol. 1, 20 May 1943.

28. Ibid., 28 May, 26 June, and 28 July 1943; Greenewalt to Compton, 12 June 1943, NARA, RG 77, entry 5, folder 441.2, Polymer. Urey also favored the slurry, homogeneous concept. Harold C. Urey to W. K. Lewis, T. C. Tolman, E. V. Murphree, E. Bright Wilson, 9 August 1943, NARA, RG 227-OSRD, S-1 Materials, box 4, Files of R. C. Tolman, folder P-9 Committee.

29. Greenewalt Notes, vol. 1, 26 June 1943.

30. The committee consisted of W. K. Lewis, R. C. Tolman, E. B. Wilson, and E. V. Murphree. NARA, RG 77, MED Decimal Files, entry 5, box 61, folder Post War Policy Committee (Chicago) 336.

31. Groves to Lewis, 9 August 1943, NARA, RG 227-OSRD, S-1 Materials, box 4, Files of R. C. Tolman, folder P-9 Committee.

32. Handwritten transcript, NARA, RG 77, entry 5, box 61, folder P-9 Reviewing Committee 334, 13 (hereafter "Handwritten transcript"). Urey's written report—Columbia Serial 100-U-M-234, 9 August 1943—is in NARA, RG 227-OSRD, S-1 Materials, box 4, Files of R. C. Tolman, folder P-9 Committee.

33. Handwritten transcript, 16.

34. Ibid., 21–22.

35. Ibid., 26.

36. Ibid., 34–35.

37. Ibid., 35.

38. Ibid., 43–44.

39. Ibid., 48.

40. Ibid., 58–60.

41. Report of the Committee on Heavy Water Work, 19 August 1943, 5-P, NARA, RG 227-OSRD, S-1 Materials, box 4, Files of R. C. Tolman, folder P-9 Committee.

42. Nightingale, *Nuclear Graphite*, 128–29. CP-5 was approved in May 1951 and went critical in February 1954. Spencer Weart, *Scientists in Power* (Cambridge: Harvard University Press, 1979), 186–89, believed the heavy-water reactor would have been a better choice and should have been chosen early on instead of the graphite-moderated designs.

43. Matthias to Groves, 19 January 1944, NARA, RG 227-OSRD, S-1 Materials, box 10, Files of R. C. Tolman, folder Tolerances in W Pile.

44. Tolman to Groves, 5 February 1944, NARA, RG 227-OSRD, S-1 Materials, box 10, Files of R. C. Tolman, folder Tolerances in W Pile.

45. How or why the first three reactors were designated by the letters B, D, and F remains a bit of a mystery. Several stories survive, some apocryphal, as to the logic of the designation system. One local legend held that it was to confuse the enemy by creating the assumption that there were at least six reactors. Suggestions that the sites were so named because of a preset alphabetical grid are not borne out by maps of the Hanford reservation. Another possibility is that when Du Pont had originally planned for six helium-cooled reactors spaced 3 miles apart, the letters A through F had been assigned to these six. Then, in May 1943, when the plan was changed to three water-cooled reactors, spaced 6 miles apart (Nichols to Williams, 27 May 1943, NARA, RG 77, entry 5, box 67, folder 400.17, MFG.-Prod.-Fabrication), the alternating sites of the original plan, B, D, and F may have been selected, thus providing the original names. While this explanation is attractive because of logic, research located no specific documentation to prove it.

46. Status Report on Hanford Engineer Works, NARA, RG 227-OSRD, S-1 Materials, MED Reports on Facilities, box 7, folder Status Report HEW as of 31 Mar 1944; ibid., folder Status Report HEW as of 30 June 1944, 2 (lessons learned); Rentenbach to Matthias, 24 January 1946, Hanford RHA, Manhattan District History (Smyth Draft), 9.8–9.9.

47. Smyth Draft, 9.21, 9.17, Hanford RHA, bk. 4, Pile Project, vol. 5, Construction, box C-309.

48. The startup procedure and the original tests are detailed in several historical "Memoranda for the File," P Department, Hagley, box 58, bk. 2, pt. 2.

49. For Fermi's presence at the startup of B reactor, see TNX History, 14 August 1945, Hagley, box 58, folder Deviations from Operating Standards During the Starting of 100-Area Piles.

50. Matthias to Groves, 3 October 1944, NARA, RG 77, MED Decimal Files, box 56, folder 319.1, Miscellaneous. Groves added a note to give credit to Wheeler for this suggestion. Although Groves later required that Matthias's letter of 3 October credit Wheeler with the suggestion of xenon as the poisoning effect, the contemporary diaries do not attribute the discovery or intuition specifically to one individual.

51. Matthias Diary, 27–30 September 1944, Hanford RHA, box A-737, 100 Area Daily Log, vol. (1 Sept. 1944 to 31 Dec. 1944).

52. Compton to Groves, 3 October 1944, NARA, RG 77, entry 5, MED Decimal Files, box 66, folder 400.12, Experiments.

53. Zinn to Compton, 3 October 1944, ibid. Although Groves criticized Zinn and Compton for not anticipating the problem, they had been hesitant to operate their reactor at the required level. Rhodes, *The Making of the Atomic Bomb*, 559.

54. Valente Diary, 27–30 September 1944, Hanford Public Reading Room, and NARA, RG 227-OSRD, S-1 Materials, box 4, Files of R. C. Tolman, folder P-9 Committee. Daniel J. Kevles, *The Physicists: The History of a Scientific Community in Modern America* (Cambridge: Harvard University Press, 1987), 328–29, also attributed the identification of poisoning to Wheeler. Kevles may have had in mind an earlier suggestion by Wheeler that fission products in their decay stages might present cross sections sufficiently large to poison the reaction. David A. Hounshell, "Du Pont and the Management of Large-Scale Research & Development," in Peter Galison and Bruce Hevly, eds., *Big Science: The Growth of Large-Scale Research* (Stanford, Calif.: Stanford University Press, 1992), 251.

55. Compton noted the satisfaction of the Du Pont folk on this point in *Atomic Quest* (New York: Oxford University Press, 1950), 193; David A. Hounshell and John Kenley Smith, *Science and Corporate Strategy: DuPont R&D, 1902–1980* (New York: Cambridge University Press, 1988), 341.

56. Memoranda for the File, P Department, Hagley, box 58, bk. 2, pt. 2.

57. Valente Diary, 18 January 1945.

58. Ibid., 9 and 18 June 1945.

59. Valente Diary, 1 May 1945; 100 Area Daily Log, vol. 3 (1 May 1945 to 18 July 1945), Hanford RHA, box A-737.

60. Greenewalt to Compton, 6 July 1943, NARA, RG 77, entry 5, box 74, folder 600.12 (P-9).

61. Oppenheimer to Thomas (of Monsanto), 16 November 1943; Thomas to Groves, 4 January 1944; Groves to Williams, 7 January 1944; Nichols to Groves, 23 September 1944; all in NARA, RG 77, entry 5, box 67, folder Materials 401.1–410.2.

62. Hewlett and Anderson, *The New World*, 634–35.

63. George C. Laurence, "Canada's Participation in Atomic Energy Development," *Bulletin of the Atomic Scientists* 3 (October 1947): 327. For CP-3, see Robert G. Sachs, ed., *The Nuclear Chain Reaction—Forty Years Later* (Chicago: University of Chicago Press, 1984), 213. CP-1 was moved to Cook County's Palos Forest Park; CP-3 was also built there. Later, Argonne Laboratory moved to its present site across the county line into Du Page County, about 5 miles due west in another forest preserve.

64. The Chicago scientists produced several early studies on what they called the new field of "nucleonics," predicting a range of peacetime applications. Perhaps the best was one authored by Enrico Fermi, James Franck, and others, published in an edition of twenty-five copies: *Prospectus on Nucleonics*, 18 November 1944, NARA, RG 77, entry 5, MED Decimal Files, 1942–1948, box 54, folder 319.1, Proposals for Research and Development in the Field of Atomic Energy. See Weart, *Scientists in Power*, 10, 38, for early anticipation of benefits from nuclear power.

65. The story of the attempts by Szilard and others at Chicago to affect the decision to drop the nuclear weapon on Japan has been told. See Rhodes, *The Making of the Atomic Bomb*, 503–10; Weart, *Scientists in Power*, 190.

Chapter Three: Contracting Atoms

1. Michael S. Sherry, *Preparing for the Next War: American Plans for Postwar Defense, 1941–1945* (New Haven: Yale University Press, 1977), 235–38.

2. Richard Hewlett and Oscar Anderson, *The New World, 1939–1946*, vol. 1 of *A History of the United States Atomic Energy Commission* (Washington, D.C.: AEC, 1962), ch. 13.

3. Ibid.

4. Ibid., ch. 14.

5. Ibid., 620–24; Richard Pfau, *No Sacrifice Too Great: The Life of Lewis L. Strauss* (Charlottesville: University Press of Virginia, 1984).

6. Atomic Energy Commission, letter to the U.S. Senate transmitting the Initial Report of the Commission, 31 January 1947, 80th Cong., 1st sess., 1947, S. Doc. 8, 7.

7. AEC, Initial Report of the Commission, 5–6; Richard G. Hewlett and Francis Duncan, *Atomic Shield, 1947–1952*, vol. 2 of *A History of the United States Atomic Energy Commission* (Washington, D.C.: AEC, 1972), 15–16.

8. Hewlett and Anderson, *The New World*, 429, 435–36; Hewlett and Duncan, *Atomic Shield*, 324; AEC, Initial Report of the Commission, 2. The evolution of the JCAE over later years is covered in Harold Green and Alan Rosenthal, *Government of the Atom* (New York: Atherton Press, 1963). On the evolution of nontechnical input from Congress, see Brian Balogh, *Chain Reaction* (New York: Columbia University Press, 1991), 73.

9. Conversation, Rodney Carlisle with Don Lewis at B reactor, Hanford, Wash., 2 June 1992.

10. David E. Lilienthal, *TVA: Democracy on the March* (New York: Harper and Brothers, 1944); William Bruce Wheeler and Michael J. McDonald, *TVA and the Tellico Dam, 1936–1979: A Bureaucratic Crisis in Post-Industrial America* (Knoxville: University of Tennessee Press, 1986), 6–7.

11. Hewlett and Duncan, *Atomic Shield*, 6.

12. The evolution of the relationship between the AEC and the weapons complex is an intricate story, only suggested here. For a fuller treatment of the efforts of Lilienthal and the Commission to simultaneously deal with Congress and with taking over the MED establishment, see Hewlett and Duncan, *Atomic Shield*, chs. 1–7.

13. For details on the Du Pont contract with MED, see Hewlett and Anderson, *The New World*, 629; Notes for the Secretary of War's Conference with Mr. Carpenter of E. I. du Pont de Nemours & Company, n.d., attached to S. L. Brown to K. D. Nichols, 3 April 1946, NARA, RG 77, entry 5, box 44, folder 161, Du Pont, 1.

14. L. R. Groves to W. S. Carpenter, 27 February 1946, and R. P. Patterson to W. S. Carpenter, 15 March 1946, NARA, RG 77, entry 5, box 44, file 161, Du Pont. See David A. Hounshell and John Kenley Smith, *Science and Corporate Strategy* (New York: Cambridge University Press, 1988), 344–45.

15. Notes for the Secretary of War's Conference with Mr. Carpenter.

16. Ibid.

17. C. E. Wilson to L. R. Groves, 28 May 1946, NARA, RG 77, entry 5, box 45, file 161, GE.

18. Hewlett and Anderson, *New World*, 629; L. R. Groves to W. S. Carpenter, 27 September 1946, NARA, RG 77, entry 5, box 44, file 161, Du Pont.

19. Hewlett and Anderson, *New World*, 629, 644–45.

20. [Hewlett], AEC 1140, "History of Expansion of AEC Production Facilities," 16 August 1963, 3, DOE Archives, RG 326, Secretariat, box 1435, folder I&P 14, History (hereafter AEC 1140); Hewlett and Duncan, *Atomic Shield*, 39–40.

21. Hewlett and Duncan, *Atomic Shield*, 40. Hewlett and Duncan show that B reactor was in reserve for general plutonium production; however, of the two products, plutonium and polonium, it was only polonium that would vanish from the stockpile because of short half-life.

22. AEC, *Fifth Semiannual Report*, January 1949 (Washington, D.C: GPO, 1949), 26; D. E. Lilienthal to Lt. Gen. L. H. Brereton, 29 April 1947, with attached Long-term Commission Agenda, NARA, RG 326, Lilienthal Subject Files, 46–50, box 9, folder Correspondence MLC, 1947; Memorandum, K. D. Nichols to L. R. Groves, 9 August 1946, with attached GE-KHK-3, "GE Nucleonics Project: Technical Program," 5 August 1946, 2, 19, DOE Archives, RG 326, Secretariat, box 1216, folder GAC; Hewlett and Duncan, *Atomic Shield*, 39.

23. Hewlett and Duncan, *Atomic Shield*, 47–48; David Alan Rosenberg, "U.S. Nuclear Stockpile, 1945–1950" *Bulletin of the Atomic Scientists*, May 1982, 26.

24. Hewlett and Duncan, *Atomic Shield*, 35, 39; GAC, Minutes of Second Meeting, 2–3 February 1947, 7, DOE Archives, RG 326, Secretariat, box 1217, folder GAC.

25. Hewlett and Duncan, *Atomic Shield*, 40; GAC, Minutes of Third Meeting, 28–30 March 1947, 26, DOE Archives, RG 326, Secretariat, box 1217, folder GAC.

26. Notes on the Meeting of the AEC and the MLC, 30 April 1947, 2, NARA, RG 326, Lilienthal Subject Files, 46–50, box 9, file Agenda and Minutes MLC-AEC 1947; Minutes of Joint MLC-AEC Meeting, 24 September 1947, 1, NARA, RG 326, Lilienthal Subject Files, 46–50, box 9, file Agenda and Minutes MLC-AEC Meetings 1947.

27. Summary of Proceedings of an Executive Meeting of the JCAE, 26 June 1947, 23, NARA, RG 128, Executive Meetings Transcripts, box 1, folder JCAE 965.

28. R. E. Nightingale, ed., *Nuclear Graphite* (New York: Academic Press, 1962), 7; W. R. Harper, *Basic Principles of Fission Reactors* (New York: Interscience Publishers, 1961), 152–53.

29. Summary of Proceedings of an Executive Meeting of the JCAE, 26 June 1947, 24.

30. GAC, Minutes of the Fifth Meeting, 28–29 July 1947, 18–19, DOE Archives, RG 326, Secretariat, box 1217, folder GAC; Hewlett and Duncan, *Atomic Shield*, 146.

31. Nightingale, *Nuclear Graphite*, 8, 16; Harper, *Basic Principles of Fission Reactors*, 153–54; Atomic Energy Office, *Final Report of the Committee Appointed by the Prime Minister to Make a Technical Evaluation of Information Relating to the Design and Operation of the Windscale Piles, and to Review the Factors Involved in the Controlled Release of Wigner Energy* (London: Her Majesty's Stationary Office, July 1958), 7, Hanford RHA, box 15200, folder Medicine, Health, and Safety 16-4 (hereafter MH&S 16-4), Reactor Safeguard Committee (1958).

32. Hewlett and Duncan, *Atomic Shield*, 146; [Hewlett], AEC 1140, 6; memorandum, Leber to Groves, 26 February 1948, Replacement and New Pile Program at Hanford, NARA, RG 77, entry 5, box 42, folder 121.2.

33. Hewlett and Duncan, *Atomic Shield*, 102, 145, 175.

34. GAC, Minutes of the Ninth Meeting, 23–25 April 1948, 2, DOE Archives, RG 326, Secretariat, box 1217, folder GAC Minutes; AEC 1140, 7–8, GAC, Minutes of the 11th Meeting, 21–23 October 1948, 9, DOE Archives, RG 326, Secretariat, box 1217, folder GAC Minutes.

35. GE, "Design and Construction History, Pile Area H," June 1952, 10, 13–16, Hanford Public Reading Room; General Business Meeting, JCAE, 26 July 1949, 1013–14, NARA, RG 128, Transcript of Executive Sessions, box 2.

36. Minutes of an Executive Meeting of the JCAE, 10 March 1949, 3–4, NARA, RG 128, Transcripts of Executive Sessions, box 2; GAC, Minutes of 12th Meeting, 3–5 February 1949, 15, DOE Archives, RG 326, Secretariat, box 1217, folder GAC Minutes.

37. Note 3–20, JCAE Executive Meeting Minutes, 10 March 1949, 5, NARA, RG 128, Transcripts of Executive Sessions, box 2; Hewlett and Duncan, *Atomic Shield*, 180.

38. JCAE Executive Meeting Minutes, 10 March 1949, 6, 10.

39. Nightingale, *Nuclear Graphite*, 8, 16. Harper, *Basic Principles of Fission Reactors*, 153–54; General Advisory Committee Minutes, 60th Meeting, 30–31 October to 1 November 1958, 21–22, 34, DOE Archives, RG 326, Secretariat, box 1387, folder 3; JCAE 10 March 1949 Executive Session Meeting Transcript, 4–5, NARA, RG 128, Transcript of Executive Sessions, box 2; HW-51188, H. E. Hanthorn, "Hanford History: Technology, Expansion and Present Efforts," 24 June 1957, 7, Hanford Public Reading Room; JCAE Executive Session Transcript, 12 March 1950, 10, NARA, RG 128, Transcripts of Executive Sessions, box 7.

40. E. R. Fleury to Area Manager, 15 May 1947, "Manhattan Engineer District History," bk.4, Hanford RHA, box C-309, folder 314.7, MED History.

41. AEC, *Seventh Semiannual Report*, January 1950 (Washington, D.C.: GPO, 1950), 127–28; JCAE, Executive Session Transcripts, 12 March 1950, 9, NARA, RG 128, Transcripts of Executive Sessions, box 7; Hanthorn, "Hanford History," 4–5. When the process tubes were bending due to the Wigner Effect, Hanford started using four-inch long slugs in the hope that they could be more readily inserted and removed from the bent tubes.

42. W. J. Williams to C. L. Wilson, 23 March 1949, Re: Dr. Bacher's Memorandum of March 3, 1949, 2, DOE Archives, RG 326, Secretariat, box 4941, folder 411.3; Information Memorandum 206, "AEC Summary Report of the Reactor Safeguard Committee," 11 August 1949, 2–3, DOE Archives, RG 326, Secretariat, box 1226, folder Reactor Development Program.

43. GAC, Minutes of the 11th Meeting, 21–23 October 1948, 12, DOE Archives, RG 326, Secretariat, box 1217, folder GAC Minutes; Information Memorandum 206, "AEC Summary Report," 6–7.

44. E. Teller to G. L. Weil (two letters), 16 August 1949, attached to Weil to W. J. Williams, Comments of Teller on Two Hanford Proposals, 24 August 1949, Hanford RHA, box 15227, folder MH&S 16-4, RSC Meetings, 1948–50; D. G. Sturges to

General Electric, 1 February 1950, Hanford RHA, box 15227, folder MH&S 16-4, RSC Meetings, 1948–50.

45. AEC 172/4, AEC Report of the 10th Meeting of the RSC, 1–3 March 1950, dated 13 April 1950, 2–3, DOE Archives, RG 326, Secretariat, box 1218, folder 337.

46. JCAE, Transcript of Executive Session, Conference with General Electric, 22 June 1950, 9–10, NARA, RG 128, Transcripts of Executive Sessions, box 6; AEC, *Seventh Semiannual Report*, 128; JCAE, Transcript of Executive Session, 12 March 1950, 13, NARA, RG 128, box 7; GAC 33rd Meeting, 5–7 February 1953, 18, DOE Archives, RG 326, job 6401, box 4932, folder 5, declassified notes.

47. GAC 33rd Meeting, 17–18. The authors examined a display of the special tools and redesigned slugs while on a tour of B reactor in June 1992.

48. GAC, Minutes of 27th Meeting, 11–13 October 1951, with attached letter of J. R. Oppenheimer to G. Dean, 13 October 1951, 5, DOE Archives, RG 326, Secretariat, box 1272, folder GAC; JCAE, Transcript of Executive Session, Conference with General Electric, 10; AEC, *Sixth Semiannual Report*, July 1949 (Washington, D.C.: GPO, 1949), 6.

49. Martin Sherwin, *A World Destroyed: The Atomic Bomb and the Grand Alliance* (New York: Knopf, 1975), 237–38.

Chapter Four: Flexible Design at Savannah River

1. JCAE Executive Session Minutes, 23 September 1949, 2–3, NARA, RG 128, Transcripts of Executive Sessions, box 2; Richard Pfau, *No Sacrifice Too Great* (Charlottesville: University of Virginia Press, 1984), 111; Gregg Herken, *The Winning Weapon: The Atomic Bomb in the Cold War, 1945–1950* (Princeton: Princeton University Press, 1981), 232–33, 301–3; statement by the President on Announcing the First Atomic Explosion in the USSR, 23 September 1949, in Philip L. Cantelon et al., eds., *The American Atom: A Documentary History of Nuclear Policies from the Discovery of Fission to the Present*, 2d ed. (Philadelphia: University of Pennsylvania Press, 1991), 112. The atmospheric monitoring procedure showed a weapon test, not a reactor accident. Truman may have used the word "explosion" as disinformation to conceal the existence of the detection method.

2. JCAE Executive Session Minutes, 17 October 1949, 110, NARA, RG 128, Transcripts of Executive Sessions, box 4; D. E. Lilienthal to F. C. Schlemmer, 3 November 1949, attached to AEC press release G-49-24, Summary of Remarks..., NARA, RG 326, Lilienthal Subject Files 46–50, box 6, folder Hanford.

3. Herken, *The Winning Weapon*, 301; Notes on Closed Meeting of the JCAE, 14 October 1949, 1, NARA, RG 128, Transcripts of Executive Sessions, box 4. See JCAE Executive Session Minutes, 23 September 1949, 4–5, NARA.

4. Lilienthal to Schlemmer, 4 November 1949.

5. JCAE, Nomination of Robert Francis LeBaron to be Chairman of the Military Liaison Committee and Expansion of the Atomic Energy Program, 18 October 1949, 39, NARA, RG 128, Transcripts of Executive Sessions, box 4.

6. Pfau, *No Sacrifice Too Great*, 112–13; Hewlett and Duncan, *Atomic Shield*, 373.

7. Lilienthal to the President, with attached Memorandum for the President from the Atomic Energy Commission, "Development of a 'Super' Bomb," 9 November

1949, 2, 4, 6, NARA, RG 128, Declassified Series 2, General Subject Files, box 60, folder Thermonuclear Program; Lilienthal, diary entry, 1 November 1949, *The Journals of David E. Lilienthal: The Atomic Energy Years, 1945–1950*, 2 vols. (New York: Harper and Row, 1964), 2:583.

8. Lilienthal to the President, 9 November 1949, with attached memorandum, 4–5, RG 128, JCAE, Declassified Series 2, General Subject Files, box 60, folder Thermonuclear Program.

9. Ibid., 4–5, 8–9.

10. J. R. Oppenheimer to Lilienthal, 30 October 1949, with attached statements by members of the GAC, DOE Archives, RG 326, Secretariat, box 1217, folder GAC Minutes.

11. W. A. Hamilton, Memorandum for the Files: "Inquiry into the Aspects of a Superweapon Program," 8 November 1949, 12, NARA, RG 128, Declassified Series 2, General Subject Files, box 60, folder Thermonuclear Program; B. McMahon to President Truman, 21 November 1949, 1–2, 7, NARA, RG 128, Transcripts of Executive Sessions, box 4.

12. Lilienthal, Diary, 1 November 1949, *Journals of David Lilienthal*, 2:583–84.

13. Pfau, *No Sacrifice Too Great*, 119, 121–22. For a crisp analysis of the debate over the "super" and Truman's decision to go ahead with its development, see Herbert F. York, *The Advisors: Oppenheimer, Teller, and the Superbomb* (Palo Alto, Calif.: Stanford University Press, 1976), ch. 4.

14. Design details of weapons, of course, remain classified; this summary of publicly available information is derived from York, *The Advisors*.

15. Transcript of JCAE Executive Meeting, "Study of the President's Decision," 31 January 1950, NARA, RG 128, Transcripts of Executive Sessions, box 4; Oppenheimer to Lilienthal, 1 February 1950, report on 19th Meeting of GAC, attached to Minutes of 19th Meeting of GAC, 31 January–1 February 1950, 3, DOE Archives, RG 326, Secretariat, box 1217, folder GAC Minutes.

16. Transcript of JCAE Executive Session, "Discussion of Supplemental Budget, Fuchs Case, British Mission, Joan Hinton Case, Hydrogen Bomb Development," 10 February 1950, 21–22, NARA, RG 128, Transcripts of Executive Sessions, box 4; Memorandum for the Record, "Weekly Conference at the AEC with Mr. Shugg and Mr. Hollis and Mr. Borden and Mr. Heller of the Joint Committee Staff," 10 February 1950, 2, NARA, RG 128, JCAE, Declassified Series 2, General Subject Files, box 60, folder Thermonuclear Program; S. T. Pike to R. LeBaron, 16 February 1950, NARA, RG 326, Lilienthal Subject Files, 46–50, box 10, folder MLC-Correspondence 1950.

17. JCAE Transcript, "Civil Defense," 17 February 1950, 11–12, NARA, RG 128, Transcripts of Executive Sessions, box 5; S. T. Pike to B. McMahon, 1 March 1950, NARA, RG 128, JCAE Declassified Series 2, General Subject Files, box 57, folder Tritium; JCAE Executive Session, 12 March 1950, 8, 11–12, NARA, RG 128, Transcripts of Executive Sessions, box 7.

18. Hewlett and Duncan, *Atomic Shield, 1947–1952*, vol. 2 of *A History of the United States Atomic Energy Commission* (Washington, D.C.: AEC, 1962), 412, 415; JCAE, "Developments in the Hydrogen Bomb," 10 March 1950, 10–11, NARA, RG 128, Transcripts of Executive Sessions, box 5.

19. JCAE, "Developments in the Hydrogen Bomb," 2–4.

20. "Bugas Panel," 25 May 1950, 4, NARA, RG 128, Transcript of Executive Sessions, box 5; JCAE Meeting, 18 May 1950, 2–3, NARA, RG 128, Transcripts of Executive Sessions, box 5; Hewlett and Duncan, *Atomic Shield*, 424; [Hewlett], AEC 1140, 15–16.

21. JCAE Minutes, 21 July 1950, NARA, RG 128, Transcripts of Executive Sessions, box 6.

22. G. E. Dean, 4 August 1950, 5, NARA, RG 128, JCAE, Transcript of Executive Sessions, box 6. Dean apologized for "jumping the gun" to get Du Pont aboard, even before Congress had actually provided the money.

23. Greenewalt commented on Du Pont's position to the JCAE in Greenewalt Testimony, 8–39, 4 August 1950, NARA, RG 128, JCAE General Correspondence, box 652, folder Savannah River Plant.

24. Chronology, 28 November 1950, NARA, RG 128, JCAE General Correspondence, box 74.

25. Dean to McMahon, 28 November 1950, NARA, RG 128, JCAE General Correspondence, box 653. Dean stated that Du Pont "carried on the active investigation of potential sites."

26. Press release on Aiken/Barnwell site, 28 November 1950, NARA, RG 128, JCAE General Correspondence, box 653. The press release provided details of the roles of the site review committee and the AEC in the site selection.

27. Exchange of correspondence between Senator McMahon and AEC chairman Gordon Dean, 25 and 26 January 1951, released 29 January 1951, NARA, JCAE General Correspondence, box 654.

28. Minutes of JCAE subcommittee on AEC Budget, 16 February 1950, NARA, RG 128. Pages 17–19 of this transcript reflect Hafstad thinking out loud about which way to go with reactor development, and show the inception of Savannah River.

29. Roger Anders, ed., *Forging the Atomic Shield: Excerpts from the Diary of Gordon E. Dean* (Chapel Hill: University of North Carolina Press, 1987), appendix.

30. Du Pont—Wilmington (DPW) document 5010, 1 April 1952, 6, 7, Hagley, series IV, box 46; see also document lists, DPW-5036-3, and "Research Programs in Support of Savannah," DPW-53-614, both in Hagley, series IV, box 46.

31. January 1951 monthly report, DPW 3210–51-1, 8, re Wigner and Wheeler, Hagley, series IV, box 45.

32. A. E. Church to H. L. Bunker, Preliminary Scope of Work, 23 October 1950, p. 3, DPW-166, Hagley, series IV, box 44.

33. Ibid.

34. Ibid.

35. Address by R. M. Evans, 16 November 1953, DPW 53-1383, Hagley, series IV, box 46.

36. Abstracts of the following studies appear in DPW-5036-3, dated 30 April 1952, Hagley, series IV, box 46: control actuator design (DPW-5173), the use of zirconium-clad thorium rods (DPW-5233), studies of the removal of scale in heat exchangers (DPW-5092), water cooling studies (DPW-5132, 5216, 5295), and studies of safeguards (DPW-6091, 6092).

37. A. E. Church to H. L. Bunker, 26 October 1950, DPW-168, Hagley, series IV, box 44.

38. Address by R. M. Evans, 16 November 1953.

39. Ibid., 6.

40. The DPW reports, cited throughout this chapter, contain much of this paper trail.

41. Address by R. M. Evans, 16 November 1953, 4–5.

42. J. E. Cole to G. H. Christensen, 18 May 1951, DPW-2236, Hagley, series IV, box 44.

43. Ibid.

44. Minutes, Meeting no. 30, Engineering Department—Explosives Department, 16 March 1951, DPW-2200-30, Hagley, series IV, box 44.

45. K. Millett to C. Nelson, 15 November 1950, DPW-189, and K. Millett to C. N. Gross, 15 November 1950, DPW-191, Hagley, series IV, box 44.

46. H. Worthington to Rogers, 17 July 1953, DPW-53-1043, Hagley, series IV, box 46.

47. On polonium, Monthly Report—Technical Division, Explosives Department, January 1951, 7, DPW-3210-51-1, Hagley, series IV, box 45, and L. Squires to C. Nelson, 19 February 1951, DPW-592, Hagley, series IV, box 45. On mixed plutonium and tritium production, R. M. Evans to C. Nelson, 31 May 1951, DPW-2318, Hagley, series IV, box 44.

48. Evans to Nelson, 31 May 1951, DPW-2318. For confirmation of plutonium instead of tritium production on the first two reactors, see J. B. Tinker to J. A. Burns, 13 June 1951, DPW-2408, Hagley, series IV, box 44.

49. Address by R. M. Evans, 16 November 1953, 7; H. W. Bellas to H. L. Bunker, 7 April 1953: "100 Area Scope of Work, Use of Enriched Slugs for Tritium Production," DPW-53-654, Hagley, series IV, box 46.

50. Evans to Nelson, 31 May 1951, DPW-2318.

51. For the nylon model, see David A. Hounshell, "Du Pont and the Management of Large-Scale Research and Development," in Peter Galison and Bruce Hevly, eds., *Big Science: The Growth of Large-Scale Research* (Stanford, Calif.: Stanford University Press, 1992), 236–61.

52. W. P. Overbeck to G. E. McMillan and W. S. Church, 7 March 1952, DPW-4809; W. P. Overbeck to L. Squires et al., 14 March 1952, DPW-4864, Hagley, series IV, box 45.

53. DPW-166, 23 October 1950, Hagley, box 44, series IV.

54. Early discussion in Atomic Energy Division, Engineering Department—Explosives Department Meeting no. 26, 16 February 1951, DPW 2200–26; record for the future in the New York Ship records, with photographs, occuping 2 ft³ in the collection; the index is in box 41, Hagley.

55. A. Michael McMahon, *The Making of A Profession: A Century of Electrical Engineering in America* (New York: AIEEE Press, 1984), 218–22.

56. Reactor order figures from John W. Johnson, *Insuring Against Disaster* (Mercer, Ga.: Mercer University Press, 1986), 69.

Chapter Five: The Arms Race Arsenal

1. Richard G. Hewlett and Francis Duncan, *Atomic Shield, 1947–1952* (Washington, D.C.: AEC, 1972), 526.
2. Ibid., 520–21; [Hewlett], AEC 1140, 15–16.
3. [Hewlett], AEC 1140, 17–18.
4. JCAE, Conference with General Electric, 22 June 1950, 2–3, 5, 8, NARA, RG 128, Transcripts of Executive Sessions, box 6.
5. JCAE, "Adequacy of Production," 21 July 1950, ibid.
6. Hewlett and Duncan, *Atomic Shield*, 532–34; [Hewlett], AEC 1140, 26–27.
7. [Hewlett], AEC 1140, 27; JCAE, "Development of the Reactor Program," 23 May 1951, 125, NARA, RG 128, Transcripts of Executive Sessions, box 10; HW-51188, H. E. Hanthorn, "Hanford History: Technology, Expansion and Present Efforts," 7, an address presented 24 June 1957 to the Hanford Laboratories Summer Institute of Nuclear Energy, Hanford Public Reading Room.
8. For the story of the decision to increase production requirements, see [Hewlett], AEC 1140, 27–45. Good examples of the debate in the JCAE over ratios, war against the USSR, and building new reactors are found in JCAE, "Aspects of the Ratio Problem," 6 June 1951, 978–1012 passim, NARA, RG 128, Transcripts of Executive Sessions, box 9.
9. [Hewlett], AEC 1140, 44–46.
10. AEC 24/22, Director of Production, "Characteristics of 'X' Reactors," 19 August 1952, 2, 5, DOE Archives, RG 326, Secretariat, box 1282, folder PLBL, Hanford (hereafter AEC 24/22).
11. See n. 40, ch. 6, below.
12. AEC 24/22; GAC Minutes, 32nd Meeting, 9–10 September 1952, 5, DOE Archives, RG 326, job 6401, box 1272, folder 1.
13. AEC 24/22, 3–4.
14. AEC 24/22, 5; K. E. Fields to C. P. Anderson, 4 November 1955, with attached press release, DOE Archives, RG 326, Secretariat, box 1285, folder PLBL, Hanford.
15. Fields to Anderson, 4 November 1955.
16. AEC, *Semi-Annual Report, July–December 1953* (Washington, D.C.: GPO, 1953), 8; AEC 24/22, 4–5.
17. HW-34834, D. G. Sturges, T. W. Hauff, and O. H. Greager, "Investigation of the KW Reactor Incident," 11 February 1955, 5, Hanford RHA, box 15327, folder Investigation of KW Incident; HW-31387, D. J. Foley, "Preliminary Assessment of Disaster Control Systems," 30 March 1954, both in Hanford RHA, box 15200, folder MH&S 16-4, RSC 1954; WASH-142, 20th Meeting of the RSC, Report to the AEC, 4 March 1953, Hanford RHA, box 11274, folder 12; J. E. Travis to General Electric, Recommendations of RSC, 22 May 1953, Hanford RHA, box 15200, folder MH&S 16-4, RSC 1953; HW-28286; A. B. Greninger to D. F. Shaw, Recommendations of Reactor Safeguard Committee, 4 June 1953, Hanford RHA, box 15200, folder MH&S 16-4, RSC 1953.
18. HW-34834, Sturges, Hauff, and Greager, "Investigation of the KW Reactor Incident," 4, 10; memorandum, D. F. Shaw to Files, 10 January 1955, K Reactor Incident, Hanford RHA, box 15237, folder PLBL 100 KE and KW 1955.

19. HW-28286, Recommendations of Reactor Safeguard Committee, 4 June 1953, 2, Hanford RHA, box 15200, folder MH&S 16-4, RSC 1953; D. F. Shaw to E. J. Bloch, Report from Advisory Committee on Reactor Safeguards, 15 April 1954, Hanford RHA, box 15200, folder MH&S 16-4 RSC 1954; HW-31741, O. H. Greager to D. F. Shaw, Reactor Safeguards Recommendations, 4 May 1954, 2, Hanford RHA, box 15200, folder MH&S 16-4, RSC 1954.

20. J. E. Travis to E. J. Bloch, 28 May 1957, Report for June 4 ACRS Meeting, Hanford RHA, box 15200, folder MH&S 16-4 RSC 1957; AEC 172/18, AEC Reports of the ACRS, 28 June 1957, with attached C. R. McCullough to K. E. Fields, 10 June 1957, Hanford, RHA, box 11237, folder 13; GAC, 60th Meeting Minutes, 30–31 October through 1 November 1958, 23, DOE Archives, RG 326, Secretariat, box 1387, folder 3; Carlisle's declassified notes.

21. McCullough to Fields, 10 June 1957.

22. Atomic Energy Office (Britain), Final Report on the Wigner Release, 5, 13, Hanford RHA, box 15200, folder MH&S 16-4, RSC 1958; E. J. Bloch to J. E. Travis, 24 October 1957, ibid., RSC 1957; Richard E. Nightingale, ed., *Nuclear Graphite* (New York: Academic Press, 1962), 493–94.

23. Bloch to Travis, 24 October 1957; memorandum, Bloch to Addressees, 31 October 1957; and O. H. Greager to A. T. Gifford, 20 November 1957, all in Hanford RHA, box 15200, folder MH&S 16-4, RSC 1957. Also C. R. McCullough to L. L. Strauss, 6 January 1958, Hanford RHA, box 11274, folder 13; Nightingale, *Nuclear Graphite*, 348.

24. HW-54853, Review of Statements by ACRS Pursuant to 20 December 1957 Meeting; 5 February 1958; and A. B. Greninger to AEC, 5 February 1958, 2, both in Hanford RHA, box 15200, folder MH&S, RSC 1958; AEC 172/22, McCullough to Strauss, 13 January 1958, attached to cover note by McCool, 4 February 1958, Hanford RHA, box 11274, folder 13.

25. HW-54853, Review of Statements by ACRS Pursuant to 20 December 1957 Meeting; Greninger to AEC, 5 February 1958.

26. Travis to Bloch, 11 February 1958, Hanford Reactor Power and Exposure Levels—ACRS Recommendations, Hanford RHA, box 15200, folder MH&S 16-4, RSC 1958; Bloch to Fields, 30 January 1958, Hanford Power Reactor Levels, Hanford RHA, box 11274, folder 13.

27. W. E. Johnson to AEC, 17 July 1958, Comments by ACRS following June 5–6 Meeting, Hanford RHA, box 15200, folder MH&S 16-4, RSC 1958; C. R. McCullough to J. A. McCone, 17 December 1958, "Power and Exposure Levels of the Hanford Reactors," Hanford RHA, box 15200, folder MH&S, RSC 1959.

28. McCullough to McCone, 17 December 1958; HW-54853, Review of Statements by ACRS Pursuant to 20 December 1957 Meeting, 5–6.

Chapter Six: Designing a Reactor for Peace and War

1. The contrast between the two engineering styles was emphasized by Dr. Dominic Monetta in a conversation with Rodney Carlisle, 17 March 1993. For further insights into Du Pont's style see David A. Hounshell, "Du Pont and the Management of Large-Scale Research and Development," in Peter Galison and Bruce Hevly,

eds., *Big Science: The Growth of Large-Scale Research* (Stanford, Calif.: Stanford University Press, 1992), 236–61.

2. Richard G. Hewlett and Jack M. Holl, *Atoms for Peace and War, 1953–1961: Eisenhower and the Atomic Energy Commission*, vol. 3 of *A History of the United States Atomic Energy Commission* (Berkeley and Los Angeles: University of California Press, 1989), 186–88, 423.

3. In a pertinent example, in 1965 the AEC appointed Milton Shaw, an aide to Rickover, to head the Division of Reactor Development and Technology. Elizabeth Rolph, *Regulation of Nuclear Power: The Case of the Light Water Reactor*, R-2104-NSF, Rand Corporation, June 1977, 36.

4. Ibid., 423. In addition to the Navy work, GE worked on research and test reactors and early boiling-water power reactors, exporting early versions to Germany, Japan, and Italy. See n. 9 below.

5. The Fermi reactor was started up in 1963, but was plagued with problems. A near-disastrous accident in October 1966, which endangered nearby Detroit, caused the reactor to be closed forever. See Walter C. Patterson, *Nuclear Power* (New York: Penguin, 1983), 135–37. Hewlett and Holl, *Atoms for Peace and War*, 200, 412–14. The full story of the effort to convert the atomic establishment to peaceful purposes is told in Hewlett and Holl's work; here, we review only the reactor side of the story.

6. Ibid., 511–12.

7. *U.S. Government Organization Manual, 1959–1960*, and *1960–1961*, as cited in Rolph, *Regulation of Nuclear Power*, 17, 20.

8. Jack Holl, Roger Anders, and Alice Buck, *The United States Civilian Nuclear Power Policy, 1954–1984: A Summary History* (Washington, D.C.: U.S. Department of Energy, 1986), 40–41.

9. As another indication of the uniqueness of N reactor, General Electric had constructed at least 25 research, test, experimental, production, and power reactors by 1964. Of these, N was the only one which was pressurized-water-cooled and graphite-moderated in design. International Atomic Energy Agency, *Directory of Nuclear Research Reactors* (Vienna: IAEA, 1989); U.S. Department of Energy, *Nuclear Reactors Built, Being Built, or Planned: 1991* (Washington, D.C.: U. S. Department of Energy, 1991).

10. W. E. Johnson, "The Hanford Convertible Reactor," address presented 7 April 1961, NARA, RG 128, JCAE General Correspondence, box 591, New Production Reactor—Hanford, vol. 3.

11. Ibid.

12. At least six different studies over the period 1958–62, using different combinations of arbitrary assumptions, can be found in NARA, RG 128, JCAE General Correspondence, boxes 590 and 591, New Production Reactor—Hanford, vols. 1 and 2.

13. W. E. Johnson to J. E. Travis, 31 March and 4 April 1958, NARA, RG 128, JCAE General Correspondence, box 590, New Production Reactor—Hanford, vol. 1; Johnson to H. M. Jackson, 24 March 1961, ibid., box 591, vol. 2.

14. Johnson to Travis, 31 March and 4 April 1958.

15. This motive for peaceful use of the atom is explored in Hewlett and Holl, *Atoms for Peace and War*.

16. B. Price to E. J. Bloch, 3 February 1959, NARA, RG 128, JCAE General Correspondence, box 590, New Production Reactor—Hanford, vol. 2.

17. Ibid.

18. A tally of the congressional votes reflecting the cross currents of political affiliation and regional loyalty was recorded by JCAE Chairman and N-reactor proponent Chet Holifield, 26 September 1962, on signing the bill authorizing the power plant, NARA, RG 128, JCAE General Correspondence, box 591, New Production Reactor—Hanford, vol. 5. Representing the private-power lobby, the Edison Electric Institute complained bitterly: see S. R. Knapp to C. Holifield, 16 May 1961, ibid., vol. 3.

19. Address by Henry Jackson, "Status and Prospects for Atomic Power Development," 28 May 1959, ibid., box 590.

20. Hewlett and Holl, *Atoms for Peace and War*, 327, 439. On the reactor at Marcoule and other European gas-cooled models, see Gilbert Malese and Robert Katz, *Thermal and Flow Design of Helium-Cooled Reactors* (La Grange Park, Ill.: American Nuclear Society, 1984), and Spencer Weart, *Scientists in Power* (Cambridge: Harvard University Press, 1979), 266.

21. R. W. Beck to C. P. Anderson, 31 March 1961, NARA, RG 128, JCAE General Correspondence, box 591, New Production Reactor—Hanford, vol. 2.

22. "Nuclear Experts Disagree on Hanford NPR Power Conversion Issue," Statement by Congressman Craig Hosmer, 10 May 1961, ibid., vol. 3.

23. Ibid.

24. O. Hurd to G. T. Seaborg and C. F. Luce, 28 November 1961, and E. Coe to G. T. Seaborg, 27 October 1961, NARA, RG 128, JCAE General Correspondence, box 591, New Production Reactor—Hanford, vol. 3.

25. C. Holifield to "Colleagues," 6 July 1962, NARA, RG 128, JCAE General Correspondence, box 371, folder NPR Determinations and Hearings.

26. F. H. Weitzel to G. T. Seaborg, 7 July 1962, ibid.

27. M. C. Leverett and R. E. Trumble, HW-83123, "Establishment of Reliability of Important Reactor Safety Systems," 2 July 1964, 5, Hanford RHA, box 15200, folder MH&S 16-4 Reactor Safeguards (July–December 1964). This document provides a detailed view of the whole GE philosophy of safety, in which planning from criteria and elementary probabilistic calculations are used.

28. Ibid., 7.

29. Ibid, 5–7. A fuller description of the growth of the probabilistic method is provided in the next chapter.

30. HW-62150 RD 3, 11 September 1961, Hanford RHA, box 22749, folder NPR System Parameters.

31. C. N. Zanger to H. H. Schipper, 4 April 1962, reporting on ACRS meeting of 30 March 1962, Hanford RHA, box 15200, folder MH&S 16-4, Reactor Safeguards (1962–1963).

32. Ibid.

33. J. Campbell to C. Holifield, 22 October 1962, NARA, RG 128, JCAE General Correspondence, box 371, folder NPR Determinations and Hearings.

34. E. J. Bauser to File, 19 November 1962, ibid.

35. HW-78085, "N-Reactor Capability Appraisal," by N-Reactor Department

Staff, Hanford Atomic Products Operation, General Electric Company, 7 July 1963, 6–7, Hanford RHA, box 9920, folder PLBL 7–3, NPR Production Costs.

36. E. F. Miller to File, "Reactor Safety: Report of ACRS Subcommittee Review of N-Reactor Startup Test Results," 24 April 1964, Hanford RHA, box 15200, folder MH&S 16-4, Reactor Safeguards (January–June 1964).

37. H. J. C. Kouts to G. T. Seaborg, 13 May 1964, Hanford RHA, box 15200, folder MH&S 16-4, Reactor Safeguards, (January–June 1964).

38. Kouts to Seaborg, 13 May 1964; "Report on Hanford N-Reactor (Phase III Operation)," NARA, RG 128, JCAE General Correspondence, box 355.

39. Reed Burn, ed., *Research, Training, Test, and Production Reactor Directory: United States of America*, 3d ed. (La Grange Park, Ill.: American Nuclear Society, 1988), 113.

40. As noted in Chapter 5, the power output represented by megawatts (thermal) for a production reactor and megawatts (electrical) for a power reactor are not equivalent, since the former indicates heat-generation levels while the latter indicates electrical output. On the average, a power reactor had a thermal megawatt rating about three times higher than its electrical megawatt rating. In 1963, six commercial nuclear generating plants were in operation: Shippingport, 60 MW(e); Dresden, 210 MW(e); Yankee-Rowe, 185 MW(e); Indian Point-1, 265 MW(e); Big Rock Point, 72 MW(e); and Humboldt, 69 MW(e). Their total electrical output was 861 MW(e), but the thermal rating of the six was about 2,580 MW(t). When N reactor came on line in 1966 with its rated 3,950 MW(t), 863 MW(e), it about equaled the total of all of the others in operation at that time. Holl, Anders, and Buck, *The United States Civilian Nuclear Power Policy*, 38–41, and Frank G. Dawson, *Nuclear Power: Development and Management of a Technology* (Seattle: University of Washington Press, 1976), 85.

Chapter Seven: Surviving Détente

1. George T. Mazuzan and J. Samuel Walker, *Controlling the Atom: The Beginnings of Nuclear Regulation, 1946–1962* (Berkeley and Los Angeles: University of California Press, 1984), ch. 14.

2. A 1965 analysis for the AEC's San Francisco Operations Office that was based on the absence of catastrophic accidents to date suggested that the likelihood of such an accident was less than 1 in 500, a figure so high that the study was not given publicity. R. J. Mulvihill, et al., *Analysis of United States Power Reactor Accident Probability*, Report no. SAN-570-3 (Los Angeles: Planning Research Corporation, 1965), 3.

3. Spencer Weart, *Nuclear Fear: A History of Images* (Cambridge: Harvard University Press, 1988), 309–28.

4. Steven L. del Sesto, *Science, Politics, and Controversy: Civilian Nuclear Power in the United States, 1946–1974* (Boulder, Colo.: Westview Press, 1979), 151. On the growing number of community groups fearful of the effluents, both waterborne and airborne, from the reactor sites at Hanford, see Michael D'Antonio, *Atomic Harvest: Hanford and the Lethal Toll of America's Nuclear Arsenal* (New York: Crown Publishers, 1993); and Michele Stenehjem Gerber, *On the Home Front: The Cold War Legacy of the Hanford Nuclear Site* (Lincoln: University of Nebraska Press, 1992).

5. S. T. Pike to B. B. Hickenlooper, 11 August 1948, and D. E. Lilienthal to J. A. Krug, 29 June 1948, both in NARA, RG 128, JCAE General Correspondence, box 356.

6. AEC, *Thirteenth Semi-Annual Report, January 1953* (Washington, D.C.: GPO, 1953), 23–26 and app. 10.

7. Glenn Lee, "The Way I See It...," *Tri-City Herald*, 30 May 1954, clipping in NARA, RG 128, JCAE General Correspondence, box 356; AEC Press Release A-342, "AEC Notifies Department of Interior It Is Withdrawing Objection to Irrigation of Wahluke Slope Secondary Zone," 30 December 1958, ibid., box 362.

8. Mazuzan and Walker, *Controlling the Atom*, ch. 14.

9. D'Antonio, *American Harvest*, documents many cases of downwinder illnesses attributed to radioactive effluents from the reactors and processing plants at Hanford.

10. DPW-53-593, 15 April 1953, Hagley, box 60, folder 5; DPW-53-977, Brinn to File, 3 July 1953, Hagley, series IV, box 46.

11. DPW-53-977, Brinn to File, 3 July 1953.

12. Ernst Frankel, *Systems Reliability and Risk Analysis* (The Hague: Martinus Nijhoff, 1984), 1. Frankel cited, among others, R. Von Mises, *Probability, Statistics, and Truth* (New York: Macmillan, 1957); E. Parzen, *Modern Probability Theory and Its Applications* (New York: John Wiley & Sons, 1960); and I. Bazovsky, *Reliability Theory and Practices* (Englewood Cliffs, N.J.: Prentice-Hall, 1961). Numerous other works published in the early and mid-1960s explored the possibilities of applying probability theories, as spelled out by mathematicians, to systems engineering problems.

13. M. C. Leverett and R. E. Trumble, "Establishment of Reliability of Important Reactor Safety Systems" (HW-83123), 5, 2 July 1964, Hanford RHA, box 9920, folder MH&S 16-4, Reactor Safeguards (July–December 1964); Douglas United Nuclear, Minutes, 97th Meeting GAC, 12–14 July 1966, DOE Archives, RG 326, Secretariat Collection, box 8008, folder GAC Minutes.

14. F. R. Farmer, "Reactor Safety and Siting: A Proposed Risk Criterion," *Nuclear Safety* 8 (1967): 539–48. Another seminal work in the field was Chauncey Starr, "Social Benefit Versus Technological Risk," *Science* 165 (1969): 1232–38.

15. Thomas Kuhn, *The Structure of Scientific Revolutions*, 2d ed. (Chicago: University of Chicago Press, 1969).

16. Edward Constant, *The Origins of the Turbojet Revolution* (Baltimore: Johns Hopkins University Press, 1980), traces a technological revolution.

17. R. E. Hollingsworth to J. T. Ramey, JCAE, 12 July 1961, NARA, RG 128, JCAE General Correspondence, box 652, folder Savannah River Plant. The Du Pont safety philosophy was detailed in a lengthy speech, A. A. Johnson, "Post Start-Up Reactor Safety," originally presented to the IAEA in May 1962; excerpts with updates in 1980 are in Hagley, box 6.

18. The semiannual reports and safety audits were in the form of memoranda. M. H. Smith to R. C. Blair, 5 October 1965, notes "ever aging" plant; see also Smith to Blair, 29 August, 1962, which adopted the note that the plants were aging as an explanation for their problems as early as 1962; range of audits: J. E. Conoway to E. P. Ruppe, 13 January 1987, all in Hagley, box 6.

19. Both Hanford and Savannah River assessments were presented in F. B. Baranowski to R. E. Hollingsworth, "Report on Hazards of Production Reactors," 26 May 1965, Hagley, box 63 (alphabetical).

20. J. W. Croach to J. D. Ellett, 20 February 1967, Hagley, box 8. In 1988 this issue would surface again in another form, as Congress and the press grew concerned over whether or not some 30 incidents (popularly dubbed the "dirty thirty") had been properly evaluated by local officials and properly communicated to headquarters. The furor began with an internal review. G. C. Ridgely to F. F. Merz, 14 August 1985, Hagley, box 63 (alphabetical).

21. A. A. Johnson, "Post Start-Up Reactor Safety." The full organizational structure is presented in DPW-75-123, "Nuclear Safety Control Procedures for the Savannah River Reactors," pt. 1, 10 May 1976, Hagley, box 6.

22. *A Forum Report*, Proceedings of the Atomic Industrial Forum, 1963, copy in Hagley, series III, box 9.

23. Remarks by C. Rogers McCullogh, ibid., p. 326.

24. G. W. Bloch to J. T. Conway, 24 November 1964, NARA, RG 128, JCAE General Correspondence, box 652, folder Savannah River Plant.

25. Undated Du Pont publicity piece (date of issue noted as March 1963), NARA, RG 128, JCAE General Correspondence, box 654.

26. R. R. Hood to J. T. Fay (with attachments), 11 April 1960, Hagley, series III, box 9. Work requests by Hood Worthington of 28 February and 14 October 1957 show the origins of Du Pont's power reactor study projects (H. Worthington to Burns, 18 July 1962, Hagley, series III, box 9). Previously, Du Pont staff had authored over 60 technical information memoranda dealing with a wide variety of power reactor matters; index and memoranda are in Technical Information Memoranda, 1957–59, ibid.

27. In the Du Pont version of the history of the HWCTR, Lombard Squires noted that a number of crucial decisions were made by the AEC rather than by Du Pont; his tone suggests that the time and small cost overrun were due to AEC-inspired delays. L. Squires to R. C. Blair, 3 September 1963, Hagley, box 64.

28. R. E. Hollingsworth to J. O. Pastore, 2 November 1964, NARA, RG 128, JCAE General Correspondence, folder Savannah River Plant.

29. The original plan to stretch out the reactor closings to minimize the impact was discussed at a Commission meeting at the inception of the closings (AEC 1132/12, AEC Reactor Operations Analysis Discussion Paper, 16 December 1963, DOE Archives, RG 326, Secretariat Collection, box 1321, folder Budget, Accounting, & Finance-2, 1965, vol. 2). The policy continued in 1966 ("Opening Statement on Special Nuclear Materials, JCAE Hearings on FY 1966 Authorization Bill, 2 February 1965," Hanford RHA, box 94661, folder Information, Policies, and Problems, Economic Impact and Conversion.)

30. Cited in AEC, *Annual Report to Congress—1964* (Washington, D.C.: GPO, 1965), 44.

31. J. Vinciguerra to J. T. Conway, 4 May 1964, NARA, RG 128, JCAE General Correspondence, box 355. At the time of the correspondence, the AEC was considering possible future conversion of F reactor to peaceful purposes.

32. Speech at disarmament conference as cited in AEC, *Annual Report to Con-*

gress—1966 (Washington, D.C.: GPO, 1967), 88. Johnson had presented cutbacks in plutonium production as a gesture of conciliation as early as his first State of the Union message, 8 January 1964 (reported in *Vital Speeches of the Day*, 15 January 1964, 196).

33. AEC, *Annual Report to Congress—1966*, 88.

34. White House press release, 20 April 1964, Hanford RHA, box 94661, folder Information, Policies and Problems, Economic Impact and Conversion.

35. On fiscal year 1964 closings, JCAE Hearings on FY 1966 Authorization Bill, 2 February 1965, ibid.

36. AEC press release, "AEC to Reduce Production of Plutonium," 18 January 1968, NARA, RG 128, JCAE General Correspondence, box 356.

37. W. S. McGuire to G. Seaborg, 7 February 1964, NARA, RG 128, JCAE General Correspondence, box 652, folder Savannah River Plant.

38. Savannah River Nuclear Study Group, "Summary Report Relative to Conversion of the Savannah River 'R' Reactor," 24 February 1965, NARA, RG 128, JCAE General Correspondence, box 654. When the report was concluded, only eleven of the original utilities were still participating in the study group. G. F. Quinn to J. T. Conway, 23 March 1965, NARA, RG 128, JCAE General Correspondence, box 652, folder Savannah River Plant (2d of 2 folders).

39. R reactor was shut down in July 1964. On its cannibalization, M. H. Smith to R. C. Blair, 29 January 1964, Hagley, box 6. The reactor was never restarted.

40. E. J. Bloch to J. T. Conway, 24 November 1964, NARA, RG 128, JCAE General Correspondence, box 652, folder Savannah River Plant.

41. R. E. Hollingsworth to J. T. Conway, 6 July 1966, NARA, RG 128, JCAE General Correspondence, box 355.

42. G. T. Seaborg to J. O. Pastore, 18 January 1968, NARA, RG 128, JCAE General Correspondence, box 356.

43. D. Russell to J. E. Webb (director of NASA), 29 July 1964, DOE Archives, RG 326, Secretariat Collection, box 1403, folder PLBL-7, 7-1-64 Savannah River.

44. AEC press release, 24 January 1967, and Burke to Conway, 20 February 1967, NARA, RG 128, JCAE General Correspondence, box 356.

45. Seaborg to Pastore, 18 January 1968, with attachments.

46. R. E. Hollingsworth to J. T. Conway, 11 August 1968; Conway to Hollingsworth, 9 August 1968; Quinn to Conway, 26 July 1968, all in NARA, RG 128, JCAE General Correspondence, box 652, folder Savannah River Plant (2d of 2 folders). Someone at the JCAE believed the AEC assurances that the leakage rate had reached a maximum were "NAIVE," and so marked Quinn's letter of 26 July.

47. G. Lee to G. T. Seaborg, 21 January 1969, NARA, RG 128, JCAE General Correspondence, box 355. Complaints about the high-handed nature of the AEC decisions were lodged at high levels: Warren Magnuson to Richard Nixon, 22 January 1970, and Warren Magnuson to Chet Holifield, 18 February 1970, NARA, RG 128, JCAE General Correspondence, box 356.

48. The five commercial power reactors that began operation in 1970 were Millstone no. 1 (Waterford, Conn.), 660 MW(e); Dresden no. 2 (Morris, Ill.), 794 MW(e); Monticello (Monticello, Minn.), 545 MW(e); Robinson (Hartsville, S.C.), 700 MW(e); and Point Beach no. 1 (Two Creeks, Wis.), 497 MW(e). Thermal ratings are not avail-

able. Jack Holl, Roger Anders, and Alice Buck, *The United States Civilian Nuclear Power Policy, 1954–1984: A Summary History* (Washington, D.C.: U.S. Department of Energy, 1986); [Hewlett], AEC 1140.

49. AEC press release E-410 and attached report, 8 November 1962, Hanford RHA, box 94661.

50. Slaton Report, 8 November 1962, Hanford RHA, box 94661.

51. The word "city" in the term "Tri-City" represented a bit of boosterism, given a total population of 55,208 in the 1960 census. While perhaps meeting the "urban area" standard by Census Bureau definition, the area was hardly metropolitan. On the incorporation of Richland, N. G. Fuller, "Hanford Real Estate History," November 1979, Hanford RHA, box A-365.

52. G. Lee to J. T. Conway, 5 May 1964, NARA, RG 128, JCAE General Correspondence, box 355.

53. Glenn Lee, "The Facts about Hanford's New Future and the Long-Range Growth of the Tri-Cities," remarks at Seattle First National Bank, 27 February 1964, NARA, RG 128, JCAE General Correspondence, box 355.

54. Lee to Conway, 5 May 1964.

55. Seaborg to Pastore, 20 January 1964, NARA, RG 128, JCAE General Correspondence, box 355.

56. P. G. Holstead and F. W. Albaugh, "The Potential for Diversification of the Hanford Area and the Tri-Cities," 15 January 1964, Hanford RHA, box 28704.

57. The notion of power conversion of a production reactor was discouraged. G. T. Seaborg to Sen. Carl Hayden, 14 December 1964, DOE Archives, RG 326, Secretariat Collection, box 1402, folder PLBL-7, Hanford, vol. 5. This idea had to be continually rebuffed, as early as the mid-fifties: D. F. Shaw to J. T. Ramey, 19 February 1957, NARA, RG 128, JCAE General Correspondence, box 534.

58. Remarks by Dr. Glenn T. Seaborg, Hanford Day at the World's Fair, 1 October 1964, Hanford RHA, box 94661, folder Information, Policies, and Problems, Economic Impact and Conversion; Seaborg to R. F. Kennedy, 24 February 1965, DOE Archives, RG 326, Secretariat Collection, box 1402, folder PLBL-7, Hanford, vol. 6. Congresswoman Catherine May supported the University of Washington proposal: C. May to G. T. Seaborg, 22 January 1966, DOE Archives, Secretariat Collection, box 1402, folder PLBL-7, Hanford, vol. 9.

59. L. B. Johnson to G. T. Seaborg, 17 April 1965, Hanford RHA, box 94661, folder Information, Policies, and Problems, Economic Impact and Conversion.

60. M. R. Schneller to H. H. Schipper, 20 November 1964, Hanford RHA, box 9920, MH&S 16-4, Reactor Safeguards (July–December 1964).

61. Ibid.

62. F. P. Baranowski to J. E. Travis, 17 July 1964, Hanford RHA, box 9920, MH&S 16-4 Reactor Safeguards (July–December 1964); Schneller to Schipper, 20 November 1964.

63. WPPSS press release, 4 November 1967, NARA, RG 128, JCAE General Correspondence, box 355.

64. G. Lee to C. Holifield, 18 August 1970, ibid., box 356.

65. F. W. Albaugh and P. G. Holstead, "Fast Fuel Test Reactor Study," 1 April 1964, Hanford RHA, box 28704.

66. *Hanford, Yesterday, Today, and Tomorrow* (public relations brochure), 3rd ed., 1977, 38, Hanford Public Reading Room.

67. Lyndon Johnson, press release, "Remarks of the President at Associated Press Luncheon, Waldorf Astoria Hotel, New York City," 20 April 1964, Hanford RHA, box 94661, folder Information, Policies, and Problems, Economic Impact of Conversion.

68. William H. McNeil, *The Pursuit of Power: Technology, Armed Forces, and Society* (Chicago: University of Chicago Press, 1982), 363–84.

Chapter Eight: Lobbying for Nuclear Pork

1. Paul Nitze, James Doughtery, and Francis X. Kane, *The Fateful Ends and Shades of SALT* (New York: Crane, Russak, 1979), ix. The quotation is from the preface, by Frank R. Barnett. Paul Nitze had represented the United States at the SALT I negotiations in 1969–74; he had served from 1961 to 1969 in Democratic administrations in Department of Defense posts.

2. INEL's acreage was later expanded by 170,000 acres to over 577,000 acres, or 891 square miles. The number of reactors is given in Susanne Miller, *Idaho National Engineering Laboratory Management Plan for Cultural Resources*, DOE/ID-10361, March 1992, 36; and in U.S. Department of Energy, Idaho Field Office, *Informal Historic Summary: Auxiliary Reactor Areas II and III, The Idaho National Engineering Laboratory*, n.d.

3. Jack Holl, Roger Anders, and Alice Buck, *The United States Civilian Nuclear Power Policy, 1954–1984* (Washington, D.C.: U.S. Department of Energy, 1986), 15.

4. Ralph R. Fullwood and Robert E. Hall, *Probabilistic Risk Assessment in the Nuclear Power Industry: Fundamentals and Applications*, Brookhaven National Laboratory (New York: Pergamon Press, 1988), 15–16; [Rodney Carlisle], *Probabilistic Risk Assessment in New Production Reactors: Background and Issues to 1991* (Washington, D.C.: U. S. Department of Energy, Office of New Production Reactors, October 1992).

5. Ralph Nader and John Abbotts, *The Menace of Atomic Energy*, 2d ed. (New York: Norton, 1979), 12.

6. Ibid., 11–12.

7. An excellent short treatment of this issue is found in Dorothy Nelkin, "Anti-Nuclear Connections, Power and Weapons," *Bulletin of the Atomic Scientists*, April 1981, 36.

8. The reader may find the acronym "NPR" somewhat confusing, since the same term was used in two separate eras, 1957–63 and 1980–92. In the first era, N reactor at Hanford was named "N" as an abbreviation for NPR. By 1980, when the term "NPR" came into use again to describe one or more planned new production reactors, N reactor was simply thought of by the single-letter designation. The term "NPR" was adopted as an official designation by congressional action in 1982. Thus, as used from 1980 to 1992, the term "NPR" referred to one or more of the notional reactors planned to provide a replacement capacity.

9. Background, 29 January 1985, 5661.1.7.2, EG&G collection, History Associates Inc., Rockville, Md. This collection, generated by EG&G as DOE's Defense Programs support contractor in the period 1982–84, was turned over to History Asso-

ciates for the duration of the preparation of this work; on completion of the writing, the materials were returned to the Department of Energy for archiving. A decimal system, rather than a system of box and file numbers, provided access, and those numbers are used here. Many duplicate copies of significant documents were labeled with different decimal numbers, so the original decimal designation will probably not be sustained after archival sorting and organization.

10. *New York Times*, 18 September 1980, 31.

11. NP-40, "NPR Chronology," 18 October 1988. This chronology, prepared by NP-40 (an office within ONPR), was provided with the EG&G collection.

12. For critics, see "Lack of Plutonium for Warheads Stirs Debate on Increasing Output," *New York Times*, 16 September 1980, 3; and William Safire, "The Plutonium Shortfall," *New York Times*, 18 September 1980, 31. The presidential order was reported in "Top Carter Aides, in Policy Shift, Back Higher Plutonium Output," *New York Times*, 27 September 1980, 3.

13. NP-40, "NPR Chronology," 18 October 1988, EG&G collection.

14. Background, 29 January 1985, 5661.1.7.2, EG&G collection.

15. Public Law 97-90, 4 December 1981, 95 Stat. 1164, sect. 102, Project 82-d-200.

16. "Reagan Plans Rise in Materials Used for Nuclear Arms," *New York Times*, 28 February 1982, 1.

17. In addition to T. Keith Glennan, the committee consisted of Frank Baranowski, Wallace B. Behnke, Manson Benedict, Robert E. Hollingsworth, Thomas H. Pigford, and William J. Howard.

18. On Peach Bottom no. 1 (Peach Bottom, Pa.) and Fort St. Vrain, (Platteville, Colo.), Gilbert Malese and Robert Katz, *Thermal and Flow Design of Helium-Cooled Reactors* (La Grange Park, Ill.: American Nuclear Society, 1984). On the inherent safety of the HTGR, Linden Blue, "Inherently Safe Nuclear Power: A Question of Resolve," (paper presented at the American Nuclear Society Topical Meeting on Safety of Next Generation Power Reactors, Seattle, Wash., May 1988); Combustion Engineering–General Atomics (CEGA), *Inherent Safety and the Modular Helium Reactor New Production Reactor* (n.p., October 1990), CEGA publication.

General Atomics, the General Dynamics division involved with reactor research and development, was purchased by Gulf Oil and became Gulf General Atomics in 1967; it became independent as General Atomics from 1973 to the present. The HTGR work for a potential new production reactor was inherited by the consortium Combustion Engineering–General Atomics (CEGA) in 1989. General Atomics bought out the Combustion Engineering interests in the consortium and retained the CEGA name. Conversation, Rodney Carlisle with Thomas A. Johnston of General Atomics, 10 June 1993.

19. *New York Times*, 18 March 1982, 20.

20. Thomas B. Cochran, *The Liquid Metal Fast Breeder Reactor: An Environmental and Economic Critique* (Baltimore, Md.: Johns Hopkins University Press, Resources for the Future, 1977), 118–25.

21. Among other independent specialists, former AEC chairman Glenn Seaborg supported the gas-cooled alternative in 1970. Ibid., 125. On General Atomics' competitive position, Frank G. Dawson, *Nuclear Power: Development and Management of a Technology* (Seattle: University of Washington Press, 1976), 141–42.

22. As indicated in n. 9 above, the EG&G collection had many duplicates of letters and reports, especially those which were politically important. Multiple photocopies were made for internal briefing purposes and for Freedom of Information Act requests, and then copies of the copies of the packet-copies were maintained. Without attempting to cite to the official copy of record, which had not yet been established, one such internal copy-packet containing many of the more significant letters is cited here for convenience sake. An early draft of the Glennan report is in this packet, NPR Status Information, with an external date of collection, 6 June 1984 (5661.1.5., document no. 261), hereafter cited as "NPR Status Information".

23. T. K. Glennan to D. P. Hodel, 15 November 1982, in NPR Status Information. See also Malese and Katz, *Thermal and Flow Design*, for background on early Idaho work.

24. "Report of the New Production Reactor Concept and Site Selection Advisory Panel" (Glennan Report), 15 November 1982, in NPR Status Information.

25. S. S. Thurmond to Hodel, 19 November 1982, in NPR Status Information.

26. Hodel to Thurmond, 15 December 1982, in NPR Status Information.

27. Management Plan, 22 December 1982, 5661.1.7.1, EG&G collection.

28. Senate Committee on Armed Services, *Environmental Consequences of the Proposed Restart of the L-Reactor at the Savannah River Plant, Aiken, S.C.: Hearings before the Committee on Armed Services*, 98th Cong., 1st sess., 9 February 1983, held in North Augusta, S.C.

29. Ibid., 4–5, 29.

30. Ibid., 46–49.

31. Ibid., 51–63, 72–73.

32. Ibid., 80–81, 93–95.

33. U.S. Department of Energy, *L-Reactor Operation, Savannah River Plant: Final Environmental Impact Statement*, DOE Rept. DOE/EIS-0108, 3 vols., May 1984.

34. J. A. McClure to Hodel, 24 January 1983, in NPR Status Information.

35. Los Alamos National Laboratory, Proposed Activities, 5661.1.7, EG&G collection.

36. Draft Project Charter, 5661.1.7.3, EG&G collection.

37. These two collections were extensively consulted in preparation of this work.

38. Hodel, memorandum to H. Roser, assistant secretary for Defense Programs, 9 August 1983, in NPR Status Information.

39. Ibid.

40. Hodel to M. Price, 16 August 1983, in NPR Status Information.

41. Thurmond to Hodel, 25 August 1983, in NPR Status Information.

42. J. Tower to Hodel, 6 September 1983, in NPR Status Information.

43. M. Price to Hodel, 13 September 1983, in NPR Status Information.

44. *Inside Energy*, 7 November 1983, 8.

45. This summary is derived from a review of several hundred letters from the public collected in the EG&G collection.

46. Hodel to members of Congress, 11 May 1984, in NPR Status Information.

47. Price to Hodel, 18 June 1984, in 5661.2, FOIA letters, EG&G collection.

48. "Contingency Plan for Light Water Graphite Reactor," 16 August 1984, 5661.1.7.4, EG&G collection.

49. C. Weinberger to R. McFarlane, 28 December 1984, 5661.7.7, EG&G collection.

50. Memoranda, 5661.1.11, EG&G collection; NP-40, "NPR Chronology," 18 October 1988, item 5, EG&G collection.

51. The larger experimental reactors to be reviewed were the Fast Flux Test Facility at Hanford; the Experimental Breeder Reactor II and the Advanced Test Reactor at Idaho; the High Flux Beam Reactor at Brookhaven; and the Oak Ridge Research Reactor and the High Flux Isotope Reactor at Oak Ridge.

52. Michael D'Antonio, *Atomic Harvest: Hanford and the Lethal Toll of America's Nuclear Arsenal* (New York: Crown Publishers, 1993), 30–53, 68–70.

53. J. H. Reuben, chair, Nez Perce Tribal Council, to M. Lawrence, DOE Richland Operations Office, 1 May 1986, Hanford RHA, box 94644, Correspondence—General, 1985; J. Weaver to members of the Subcommittee on General Oversight, Northwest Power and Forest Management, Committee on Interior and Insular Affairs, 18 June 1986, Hanford RHA, box 94644, Congressional Correspondence—1986.

54. The design and safety reviews were issued as DOE publications DOE/EH-0017 and DOE/EH-0015.

55. U.S. General Accounting Office, *Nuclear Science: Issues Associated with Completing WNP-1 as a Defense Materials Production Reactor: Report to the Honorable Brock Adams, U.S. Senate,* 21 September 1988; U.S. General Accounting Office, *Nuclear Safety: Comparison of DOE's Hanford N-Reactor with the Chernobyl Reactor: Briefing Report to Congressional Requesters,* 1986; House Committee on Interior and Insular Affairs, Subcommittee on General Oversight, Northwest Power and Forest Management, *N-Reactor at Hanford Reservation, Washington: Safety and Environmental Concerns: Oversight Hearings before the Subcommittee,* 99th Cong., 2d sess., hearings held on 19 May 1986 in Portland, Ore., and on 16 June 1986 in Washington, D.C.; "U.S. to Shut Down Hanford Reactor for Safety Repairs," *Philadelphia Inquirer,* 13 December 1986.

56. N was placed on "cold standby" in mid-1988. Although the terminology about the status was changed, in effect the reactor remained closed after January 1987.

57. National Academy of Sciences, National Research Council, Commission on Physical Sciences, Mathematics and Resources, and Commission on Engineering and Technical Systems, *Safety Issues at the Defense Production Reactors* (Washington, D.C.: National Academy Press, 1987), xvii, xix, xx.

58. Ibid., 5–7.

59. C. G. Halstead, "New Production Reactor Assessment Update," February 1987, 5661.1.11.4, EG&G collection; Karen Fitzgerald, "Nuclear Weapons Reactors: Too Hot to Handle?" *IEEE Spectrum* (June 1989), 41.

60. "WNP-1 Chronology of Events" (a listing of various WNP-1 advocacy efforts through 1986 and 1987), document 0928871, Hanford RHA, box 103367, folder WNP-1 Conversion—1987.

61. *Inside Energy,* 11 January 1988, 8.

62. U.S. Department of Energy, *Assessment of Candidate Reactor Technologies for the New Production Reactor: A Report of the Energy Research Advisory Board,* DOE/S-0064, July 1988; U.S. Department of Energy, *Site Evaluation Report for the New Pro-*

duction Reactor, DOE/DP-0053, July 1988 (with attachments 0054–0056); Senate Committee on Armed Services, Subcommittee on Strategic Forces and Nuclear Deterrence, *New Production Reactor Acquisition Strategy: Hearings before the Subcommittee*, 100th Cong., 2d sess., 5 October 1988, 22–24.

63. John Ahearne, "Fixing the Nation's Nuclear-Weapons Reactors," *Technology Review* 92 (July 1989): 5. Ahearne was the chairman of DOE's Advisory Committee on Nuclear Safety and a former head of the NRC. His views on restart problems were also noted in two letters to Sen. Sam Nunn of 14 December 1988, included in Senate Committee on Governmental Affairs, *Oversight of Cleanup and Modernization Proposals for DOE's Weapons Production Complex: Hearings before the Committee*, 101st Cong., 1st sess., 25–26 January 1989, 287ff.

64. An excellent and thoughtful analysis of the various sources of the October–December 1988 media coverage is William Lanouette, "Tritium and the *Times*—How the Nuclear Weapons Production Scandal Became a National Story" (Research Paper R-1, Joan Shoenstein Barone Center for Press, Politics, and Public Policy, John F. Kennedy School of Government, Harvard University, May 1990). See also D'Antonio, *Atomic Harvest*, 249–57.

65. *United States Department of Energy Nuclear Weapons Complex Modernization Report* (the "2010 Report"), DOE/S-9006836, Report by the President to Congress, December 1988.

66. Senate Committee on Armed Services, *New Production Reactor Acquisition Strategy*.

67. J. Carson Mark et al., *The Tritium Factor* (Washington, D.C.: Nuclear Control Institute and American Academy of Arts and Sciences, December 1988), 138–40.

68. DOE Press Release, 19 January 1989, R-89-004.

Chapter Nine: Managing Nuclear Options

1. J. McClure and M. Hatfield to James Watkins, 18 May 1988, copy from "NPR Master File," doc. no. 2034, provided to author by ONPR; Sid Morrison, "Politics Determined Reactor Sites, Hanford Supporters Say," *Richland (Wash.) Tri-City Herald*, reprinted in *Augusta (Ga.) Chronicle*, 4 August 1988; "Conversion Backers Haven't Given Up Hope," and "Congressional Split Rules Out Hanford," *Tri-City Herald*, 4 August 1988.

2. Thomas J. Peters and Robert H. Waterman, Jr., *In Search of Excellence: Lessons from America's Best-Run Companies* (New York: Warner Books, 1982).

3. Ibid., 82–85.

4. "Nuclear Reactor Plan Meets Strong Opposition in South Carolina," *New York Times*, 8 December 1988, B-18. One journalistic author, a mildly antinuclear critic, applied the language of cultural analysis to the Hanford site and its employees, with much the same "time-warp" thesis; see Paul Rogat Loeb, *Nuclear Culture* (New York: Coward, McCann, 1982; Philadelphia: New Society Publishers, 1986).

5. The gulf had become apparent to many observers. "Nuclear Reactor Plan Meets Strong Opposition in South Carolina," *New York Times*, 8 December 1988, B-18.

6. Greg Esterbrook, "Radio Free Watkins," *Washington Post Magazine*, 18 February 1990, 33–34.

7. "Environmental Organizations Give DOE Weapons Program Failing Grade," *Inside Energy*, 22 April 1991.

8. Harvey M. Sapolsky, *The Polaris System Development* (Cambridge: Harvard University Press, 1972).

9. Richard Hewlett and Francis Duncan, *Nuclear Navy, 1946–1962* (Chicago: University of Chicago Press, 1974); Francis Duncan, *Rickover and the Nuclear Navy: The Discipline of Technology* (Annapolis, Md.: Naval Institute Press, 1990); Theodore Rockwell, *The Rickover Effect: How One Man Made a Difference* (Annapolis, Md.: Naval Institute Press, 1992).

10. Rodney Carlisle, *Powder and Propellants* (Washington, D.C.: GPO, 1991).

11. The most comprehensive critique, which itself summed up the history of prior criticisms and studies, was by Booz, Allen & Hamilton, "Review of Navy R&D Management, 1946–1973," 1 June 1976, Navy Laboratory/Center Coordinating Group Archives, White Oak, Md.

12. Thomas C. Hone, *Power and Change: The Administrative History of the Office of the Chief of Naval Operations, 1946–1986* (Washington, D.C.: Naval Historical Center, 1989), 121.

13. Through the 1960s and 1970s, officers entering Rickover's program received an intensive reactor course which he had set up at Mare Island and then at Orlando, Florida. Literally hundreds of alumni of these early programs moved into careers in nuclear engineering in government and the private sector and were found throughout DOE and contractor staff. The training is described in *Naval Nuclear Propulsion Program, 1972–1973: Hearings before the Joint Committee on Atomic Energy*, 92nd Cong., 2d sess., and 93rd Cong., 1st sess., testimony of 8 February 1972, pt. 1, 16–17. On the spread of Rickover-trained personnel, see Norman Polmar and Thomas B. Allen, *Rickover* (New York: Simon and Schuster, 1982), 300–303.

14. "Congressional Study Challenges Federal Use of Private Contractors," *New York Times*, 12 September 1991.

15. "Savannah River Account Fudged to Overcome Shortfalls, IG Says," *Nucleonics Week*, 9 May 1991.

16. Herrington's undersecretary, Joseph Salgado, had admitted the lack of technical expertise in the face of the National Academy of Science report criticizing the department on this score. Salgado publicly discussed the reliance upon contractor expertise in a press briefing, "Department of Energy Response to the NAS Report," 29 October 1987, 20, Hanford RHA, box 103367, folder N Reactor—NAS Report. The NAS report itself stated: "The Department, both at headquarters and in its field organizations, has relied almost entirely on its contractors to identify safety concerns and to recommend appropriate actions, in part because the imbalance in technical capabilities and experience between the contractors and DOE staff is of sufficient magnitude to preclude DOE from properly performing its audit function." National Academy of Science, National Research Council, Commission on Physical Sciences, Mathematics and Resources, and Commission on Engineering and Technical Systems, *Safety Issues at the Defense Production Reactors* (Washington, D.C.: National Academy Press, 1987), xix.

17. Harold Orlans, *Contracting for Atoms* (Washington, D.C.: The Brookings Institution, 1967), 11.

18. "Arms Plant Firms Hid Cost Overruns by Shifting Funds: Report by Energy Dept.: Operators at Savannah River Used a Building Account to Conceal Spending," *New York Times*, 7 May 1991, A1; "Nuclear Arms Plant Contractors Hid Overruns; DOE Audit Finds Funds Were Illegally Shifted Among Accounts," *Washington Post*, 8 May 1991; Mike Synar: "U.S. Opens Inquiry on Nuclear Weapon Plant," *New York Times*, 9 May 1991, A26.

19. The lack of funding for maintenance and updating of facilities was a complaint of Du Pont at Savannah River before the company declined to stay on there after 1988. "Du Pont Relinquishes National Plant," *New York Times*, 1 April 1989.

20. On the culture of complacency, "Aging N Waste Tank Sparks Fears," *Portland (Ore.) Oregonian*, 8 September 1991; on the culture of neglect, David Albright, "Tritium Supply Doesn't Warrant NPR," *Idaho Falls Register*, 1 December 1989. Such phrasing became common in news items through the period. The difficulty of managing the large GOCO contractors had persisted from the beginning: see Orlans, *Contracting for Atoms*, esp. 11–41. The relationship between the contractors and DOE through the 1980s was a continuing subject of detailed press attention. One fair summary: "Decade of Criticism Belts SRS," *Augusta Chronicle*, 31 December 1989.

21. "Watkins Restates Need for Higher Pay to Attract Superior Personnel," *Inside Energy*, 2 April 1990, 12b. That critics' expectations were not met is revealed in their "report cards" on Watkins: see, e.g., "Environmental Organizations Give DOE Weapons Program Failing Grade," *Inside Energy*, 22 April 1991.

22. U.S. Department of Energy, Secretary of Energy Notice SEN-3-89, 15 March 1989; "Transitions," *Environmental Forum*, March/April 1990, 56; "Aging N Waste Tank Sparks Fears," *Portland Oregonian*, 8 September 1991 (on local officials bringing new vigor to cleanup efforts); "Watkins Links Contractors' Bonuses to Environmental Performance," *Inside Energy*, 3 July 1989, 7; U.S. Department of Energy, DOE News Release R-89-068, 27 June 1989; Jay Olshansky and R. Gary Williams, "Culture Shock at the Weapons Complex," *Bulletin of the Atomic Scientists*, September 1990, 29–33; U.S. Department of Energy, Secretary of Energy Notice SEN-13-89, 11 November 1989.

23. "Savannah River Account Fudged"; "Arms Plant Firms Hid Cost Overruns." The funds had gone to construction of needed facilities, not to personal uses. Later investigation cleared the participants in the fund transfer of charges of impropriety.

24. "Department's Woes Blur Watkins' Bold Vision," *The State* (Columbia, S.C.), 8 May 1991.

25. "Retired Admiral Named New Manager at SRS," *Aiken (S.C.) Standard*, 17 July 1991.

26. DOE Notice N1100.21, 16 November 1988.

27. DOE News Release R-89-071, 3 July 1989.

28. "Monetta Says Light Water Tritium Target Proven in Recent NPR Testing," *Inside Energy*, 20 May 1991. This article reviewed the effort to reduce technopolitical bias.

29. For a fuller treatment of Browning's administrative style and the Assistant Management Board, see Carlisle, *Powder and Propellants*, 206–7. Many of the observations of Monetta's style and vocabulary derive from Carlisle's attendance at more than ten management meetings during the period 1989–92.

30. Monetta, in response to a question at a conference with graduate students from Virginia Polytechnic Institute on 6 September 1991 at the River Inn, Washington, D.C., vividly described the blending of the separate institutional cultures into a new, diverse culture at ONPR.

31. These points were made explicitly at a 7 December 1990 meeting between Carlisle and Monetta at the River Inn, Washington, D.C.

32. Conversation between Carlisle and Monetta, 7 December 1990. Monetta's managerial methods were further documented in a series of monographs published by Virginia Polytechnic Institute as reports for DOE, together with several M.S. theses which resulted from the same project. See Harold A. Kurstedt and D. Scott Sink, *Research and Development of Post-Classical Management Tools for Government Program Offices*, DOE/NP/00119-3, vols. 2 and 3 (Blacksburg: Virginia Polytechnic Institute, February 1993); Eileen Morton Van Aken, "A Multiple Case Study on the Information System to Support Self-Managing Teams" (M.S. thesis, Virginia Polytechnic Institute, 1991); Timothy Kotnour, "Design, Development, and Testing of an Automated Knowledge-Acquisition Tool to Aid Problem Solving, Decision Making, and Planning" (M.S. thesis, Virginia Polytechnic Institute, 1992).

33. The Rickover quotations were distributed at an off-site meeting at Piney Point, Maryland, on 22 June 1991; personal observation by Carlisle, in attendance.

34. "Monetta Says Light Water Tritium Target Proven in Recent NPR Testing," *Inside Energy*, 20 May 1991.

35. "DOE Releases Draft Environmental Study for New Production Reactor," *Inside Energy*, 15 April 1991.

36. "NPR Hearings," *Idaho Mountain Express* (Ketchum, Idaho), 17 April 1991.

37. Two of the participants on the NAS panel who served as senior consultants to ONPR were George Apostolakis, University of California, and Neil Todreas, Massachusetts Institute of Technology.

38. General Atomics and friends of the firm at INEL had presented convincing arguments regarding inherent safety and electric generation prospects consistently through the 1980s, winning some support even among academics and members of the Union of Concerned Scientists. "Experts Call Reactor Design 'Immune' to Disaster," *New York Times*, 15 November 1988, C-1; "How to Build a Safe Reactor," *Time*, 29 April 1991, 58.

39. "Ebasco Touts Water-Cooled Nuclear Reactor," *Aiken (S.C.) Standard*, 22 April 1991; "Ebasco Cites Advantages of Heavy Water Reactors," *Aiken Standard*, 23 April 1991.

40. For a fuller discussion of this issue, see [Rodney Carlisle], *Probabilistic Risk Assessment in New Production Reactors: Background and Issues to 1991* (Washington, D.C.: ONPR, October 1992).

41. "Citing Friendship with Ebasco Lobbyist, Gault Will Avoid NPR Debate," *Inside Energy*, 25 November 1991. Questions about a trip Watkins and Polly Gault, his assistant, made to Alaska with the Ebasco lobbyist surfaced when a dismissed DOE employee filed a complaint under the Whistleblower Protection Act: see "Ex-NPR Aide Says DOE Illegally Fired Him," *Inside Energy*, 16 December 1991; "Lobbyist's Travel Favors Raising Questions: Energy Secretary, Guests Accepted Trip to Alaska," *Washington Post*, 10 December 1991, A-11.

42. "Panel Picks Reactor Location in Advance: Hill Pressure to Build Arms Facility in S. Carolina Assailed as Premature by Environmentalists," *Washington Post*, 10 May 1991. Butler Derrick had prepared the resolution, which Spratt got through the committee: "Lawmakers Devise Bill Aiding SRS," *The State* (Columbia, S.C.), 9 May 1991.

43. "Political Might May Decide NPR Location," *Idaho Falls Post Register*, 12 May 1991.

44. "K Reactor Problems Hurt Rating for Westinghouse at Savannah River," *Weapons Complex Monitor*, 22 June 1992; Steve Piacente, "Energy's Angst," *Government Executive*, June 1992, 48–51. Some estimates put the accumulated expenses on the K reactor remodeling as approaching $2 billion. The K restart plan, however, was canceled in early 1993 by Watkins' successor, Hazel O'Leary; K reactor was closed permanently. "Savannah River Reactor to Stay Shut; 1200 Layoffs Set," *Washington Post*, 31 March 1993.

45. An early prediction of the collapse of the system was greeted with critical disdain; see Andrei Amalrik, *Will the Soviet Union Survive until 1984?* (New York: Harper & Row, 1970).

46. The connection between the cut in weapons and the NPR delay was noted by Keith Schneider, "Weapons Cuts Lead White House to Question Plan for New A-Plant," *New York Times*, 4 October 1991, A-1.

47. "Energy Department Defends Nuclear Arms Program," *New York Times*, 19 April 1991; "Watkins Admits Arms Cuts Diminish Urgency of K-Reactor Restart," *Inside Energy*, 22 April 1991; "Global vs. Local: International Debate over Arms Race Could Shape Local Economy," *Idaho Falls Post Register*, 12 May 1991.

48. "Wimp Factor May Be Key to Miller's Stand on NPR," *Aiken (S.C.) Standard*, 21 June 1991.

49. "Cold War's End Reshapes Jobs," *Augusta (Ga.) Chronicle*, 10 May 1992; "Changing Missions—SRS Heads the Technology Era," *Aiken Standard*, 28 May 1992.

50. The postponed decision and Monetta's resignation were covered in the South Carolina, Idaho, and northwestern press, as well as in Washington (*Washington Post*, 2 November 1991; Associated Press, 6 November 1991). Newsletter coverage was also thorough: e.g., *Energy Daily*, 18 November 1991; *Weapons Complex Monitor*, 25 November 1991.

51. Sam Nunn and J. James Exon to Watkins, 14 November 1991, cited in "Nunn, Exon Slam DOE in Weapons Complex Debate," *Energy Daily*, 18 November 1991.

52. *Washington Post*, 5 November 1991; Carlisle, interview with Tom Hendrickson, 2 April 1992, Washington, D.C.

53. Ibid.

54. *Inside Energy*, 14 September 1992.

55. "DOE's Monetta Lands Position at Department of Defense," *Energy Daily*, 13 December 1991; Steve Piacente, "Energy's Angst," *Government Executive*, June 1992, 51 (on Tuck).

56. This policy had been anticipated by Watkins earlier in the year. "Energy Department Defends Nuclear Arms Program," *New York Times*, 19 April 1991; "Watkins Admits Arms Cuts Diminish Urgency of K-Reactor Restart."

57. Holly Idelson, "Nuclear Weapons Complex Braces for Overhaul," *Congressional Quarterly*, 25 April 1992, 1066–73.

Conclusion

1. Journalistic commentators were well aware of the likely resurgence of the political battle at the end of the process: see "Major Fight Seen on Bomb Reactors Proposed by U.S.: No Consensus on Need: Debate over Nuclear Arsenal Centers on Technological and Political Concerns," *New York Times*, 5 December 1988, A-1. The fear that Senator McClure would cut the DOE budget if the program did not stay alive is reported in "DOE May Not Build Plutonium Plant," *Washington Post*, 28 November 1989, A-6. McClure chose not to run for reelection in 1990, and the impact of his decision was clear, as is seen in a headline in *Nucleonics Week*, 11 January 1990, "McClure Retirement Could Jeopardize MHTGR Prospects." The continuing support of Strom Thurmond and Butler Derrick for the South Carolina site was apparent during the ONPR effort as they backed the Spratt resolution and as they continued to argue for Savannah River; see "East Not Affecting NPR Fate," *Aiken (S.C.) Herald*, 2 December 1989. Rep. Sid Morrison continued to champion the Hanford site: "Hanford Pitched for Reactor," *Aiken (S.C.) Standard*, 11 April 1991.

2. Ted Gold, Hicks & Associates, Inc., and Rich Wagner, Kaman Corp., "Long Shadows and Virtual Swords: Managing Defense Resources in the Changing Security Environment," June 1990, as cited in Charles B. Cochrane, "DoD's New Acquisition Approach: Myth or Reality?" *Program Manager*, July–August 1992, 38–45.

BIBLIOGRAPHIC ESSAY

Archival documents available for this topic are daunting in quantity. We have tapped a part of them, trying to concentrate on materials already declassified. Although we and the research assistants on the project had security clearances that allowed us to work in files that had not yet been declassified, we focused on works available to scholars without such access. Even with this self-imposed limitation we found far more than we could possibly use.

At the National Archives and Records Administration (NARA), we worked with five record groups: RG 128, RG 77, RG 227, RG 326, and RG 359. At the Department of Energy, we worked with records held by the Office of the Historian for the DOE Archives. At Hanford, we reviewed about 27 cubic feet of records in the Records Holding Area (RHA) and went through the open shelves in the published and collected documents in the Public Reading Room. At the Hagley Museum in Wilmington, Delaware, we worked with 60 cubic feet of Du Pont papers in Accession 1957.

In addition, several other collections were made available to us in the course of the project. We inherited some 11 linear feet of EG&G records of the early Defense Program office involved in reactor planning, DP-132, covering the period 1982–84, which we turned over to DOE's chief historian on completion of the project. In addition, we had temporary possession of some 65 linear shelf feet of Argonne National Laboratory, Germantown, records for the period when that office had served as the office support contractor to DP-132 and to the Office of New Production Reactors, 1985–89. Those records were returned to ONPR when it went out of business early in 1993, and they became part of the DOE archives for ONPR. During the period 1989–91, we were invited to review the office files of the various offices of ONPR and to photocopy current documents. We gathered some 10 cubic feet of documents in those searches, and that collection of ONPR documents was also turned over to the DOE chief historian on completion of the project.

While we thus reviewed well over 500 cubic feet of documents, the number we selected for direct work on the project represented about 10 cubic feet. We used about half of those, and they are cited in our notes.

In general, separate collections were useful for separate periods. The records of the Office of Scientific Research and Development and of the Manhattan Engineer District in NARA RG 77, Office of the Chief of Engineers, and RG 227, Office of Scientific Research and Development, provided the best material for the World War II period. RG 128, Joint Committee on Atomic Energy; RG 326, Atomic Energy Commission; RG 359, Office of Science and Technology; and DOE records were strong for the Atomic Energy Commission period from 1946 to the 1960s. The Records Holding Area at Hanford was useful both for the early period, the development of N reactor there, and later operation through the 1970s and 1980s. The Hagley collection provided insight and documentation for the building of Savannah River reactors, as well as for operations through 1988. The EG&G and Argonne collection from DP-132 were helpful in shedding light on the political gridlock of the 1980s. Documents from ONPR and discussions with participants were useful in the chapter dealing with the efforts to select a design and contractor, 1989–92.

Carlisle had the opportunity to meet with managers of ONPR a number of times and to attend an off-site meeting at Piney Point, Maryland. Through these direct observations and meetings, he witnessed the effort to establish some of the office's managerial styles firsthand. A few formal and informal oral history interviews and direct observations thus supplemented the documentary record. In addition, newsletters, press releases, Secretary of Energy Notices, and other public documents fleshed out the unfolding story of ONPR. We had the experience, somewhat rare for historians, of writing about an institution which we could observe firsthand in our own times.

Technical books and reports detail the workings of reactors and the nature of risk analysis, resulting in a large "gray literature" of corporate reports, advisory board publications, and government documents which are essential to understanding the technopolitical controversies among advocates of different technical systems. We have treated this body of reports as primary documentation, and it is presented throughout the note citations. While the records we reviewed contained many photocopied and low-circulation reports as well as published reports, we have tried to identify materials available in libraries, rather than only in archives. We have reviewed and quoted from many newspapers and newsletters, and these are cited only in the notes.

The study of production reactor history draws on and touches several distinct bodies of published literature. There is solid work on the organization of the AEC and its successor agencies, written for the most part by the official historians of the agency, together with several well-documented works on the history of the development of the atomic bomb. Another literature reviews national and international atomic weapons policy as an aspect of strategic history. Although the focus of this book is on the reactors used to produce strategic materials for nuclear weapons, the public controversies surrounding their better-known and younger cousins, the electrical power reactors, have spawned several shelves of works examining issues of environmental and safety risk in connection with nuclear reactors. In recent decades, the literature of management has proliferated with the simultaneous growth of graduate business programs and the popularizing of management theory.

These separate literatures each have their own controversies and divisions between devoted advocates of opposing lines of thought, often leading to hasty or polemical works. However, some authors who strenuously defend particular positions do so with rigorous scholarship and extensive research, providing access not only to their line of argument but to a rich collection of facts. Here, we evaluate a few of the strongest works in the various fields, concentrating on those that provide the best thinking, analysis, and detail.

The bibliography of histories, memoirs, and diaries documenting the history of the atomic weapon is extensive. The best memoirs are Leslie Groves, *Now It Can Be Told: The Story of the Manhattan Project* (New York: Harper, 1962), and Arthur Holly Compton, *Atomic Quest: A Personal Narrative* (New York: Oxford University Press, 1950); a good diary is that of David Lilienthal, *The Journals of David E. Lilienthal: The Atomic Energy Years, 1945–1950.* (New York: Harper and Row, 1964.) An excellent unpublished diary used extensively in this work is that of Crawford Greenewalt, maintained at both the DOE Archives and at the Hagley Archives in Wilmington, Delaware.

The best scholarly works are Richard G. Hewlett and Oscar E. Anderson, Jr., *The New World, 1939/1946*, vol. 1 of *A History of the United States Atomic Energy Commission* (University Park: Pennsylvania State University Press, 1962); Vincent Jones, *Manhattan: The Army and the Atomic Bomb* (Washington D.C: Center for Military History, U.S. Army, 1985); and Richard Rhodes, *The Making of the Atomic Bomb* (New York: Simon and Schuster, 1986). A work which provides background on the French, British, German, and the American reactor decisions in World War II and which explores some of the cultural tensions in the Manhattan project is Spencer Weart, *Scientists in Power* (Cambridge: Harvard University Press, 1979). The best work on the German project is Mark Walker, *German National Socialism and the Quest for Nuclear Power, 1939–1949* (New York: Cambridge University Press, 1989), which shows that German culture led to plans for later eventual development of nuclear energy as a power source, not a weapon. A popular and thoroughly researched work which suggests a more conscious decision not to pursue the nuclear weapon is Thomas Powers, *Heisenberg's War: The Secret History of the German Bomb* (New York: Knopf, 1993).

The literature of organizational culture has developed on two fronts since the early 1970s. On the historical side, Louis Galambos has written two historiographic articles which describe what he calls the "organizational synthesis" in historical writing: "The Emerging Organizational Synthesis in Modern American History," *Business History Review* 44 (Autumn 1970): 279–90, and "Technology, Political Economy, and Professionalization: Central Themes of the Organizational Synthesis," *Business History Review* 57 (Winter 1983): 471–93. On the popular, business-advice side, the "corporate culture" thesis pervaded management thinking beginning in 1982. Perhaps the most articulate and substantial spokesman of that school of management thought is Edgar H. Schein; see his "Corporate Culture," *Sloan Management School Review* (Winter 1984). The most popular work was Thomas J. Peters and Robert H. Waterman, Jr., *In Search of Excellence: Lessons from America's Best-Run Companies* (New York: Warner Books, 1982). Early spokesmen of the view that strong companies had "strong cultures" were Terrence E. Deal and Allen A. Kennedy, *Corporate Cultures: The Rites and Rituals of Corporate Life* (Reading, Mass.: Addison Wesley, 1982). A collection of essays directed at the same problem is Ralph Kilmann et al., eds., *Gaining Control of the Corporate Culture* (San Francisco: Jossey-Bass, 1985). A 1990 addition to the literature addressed some of these issues in governmental agencies: James Q. Wilson, *Bureaucracy: What Government Agencies*

Do and Why They Do It (New York: Basic Books, 1990). In social science, the term "culture" is usually associated with a whole society. With apologies to purists, the word "culture" is used in the less rigorous way adopted in the language of business management throughout this work.

There are several works which provide a good understanding of Du Pont's specific corporate culture. The company history is told in Max Dorian, *The du Ponts: From Gunpowder to Nylon* (Boston: Little, Brown, 1962), in a fairly traditional corporate history; the corporate style of research is elaborated in David A. Hounshell and John Kenly Smith, *Science and Corporate Strategy: Du Pont R&D, 1902–1980* (New York: Cambridge University Press, 1988). A very pointed description of the Du Pont research method as it evolved with regard to nylon and its application at Hanford is made in David A. Hounshell, "Du Pont and the Management of Large-Scale Research & Development," in Peter Galison and Bruce Hevly, eds., *Big Science: The Growth of Large-Scale Research* (Stanford, Calif.: Stanford University Press, 1992), 236–61. An early treatment of the emergence of the nuclear government-owned, contractor-operated (GOCO) establishments is Harold Orlans, *Contracting for Atoms* (Washington, D.C.: The Brookings Institution, 1967).

For the systems approach of electrical engineers, see Thomas P. Hughes, *Networks of Power* (Baltimore: Johns Hopkins University Press, 1983). Hughes further explores the systems methods of electrical innovators and engineers in *American Genesis: A Century of Invention and Technological Enthusiasm* (New York: Viking Penguin, 1989). A. Michael McMahon, in *The Making of a Profession: A Century of Electrical Engineering in America* (New York: IEEE Press, 1984), shows how electrical engineers came to dominate the organizational world of nucleonics.

As to the culture and style of the physicists themselves, Robert Jungk, in *Brighter Than a Thousand Suns* (New York: Harcourt Brace, 1958), provides a readable description of the international fraternity in physics in the interwar years; a more scholarly treatment is found in Richard Rhodes, *The Making of the Atomic Bomb*. Donald Kevles treats the lives and work of many of the players in *The Physicists: The History of a Scientist Community in Modern America* (Cambridge: Harvard University Press, 1987). For details of the early precedents for academic physicists laboring as craftsmen, see J. L. Heilbron and Robert W. Seidel, *Lawrence and His Laboratory: A History of the Lawrence Berkeley Lab*, vol. 1 (Berkeley and Los Angeles: University of California Press, 1989).

The details of the Navy's separate corporate culture can be documented through a host of good works. Harvey M. Sapolsky, *The Polaris System Development* (Cambridge: Harvard University Press, 1972), provides a look at the origins of the management styles of the Special Projects Office. Theodore Rockwell, *The Rickover Effect: How One Man Made a Difference* (Annapolis: Naval Institute Press, 1992), discusses Rickover's influence. For further description of Rickover's management approaches, see Francis Duncan, *Rickover and the Nuclear Navy: The Discipline of Technology* (Annapolis: Naval Institute Press, 1990). Rickover's specific management innovations are discussed in Richard Hewlett and Francis Duncan, *Nuclear Navy, 1946–1962* (Chicago: University of Chicago Press, 1974). For Rickover anecdotes, see Elmo Zumwalt, *On Watch: A Memoir* (New York: Quadrangle/Time Books, 1976). For a description of Rickover's methods of contractor management, see Eugene Lewis,

Public Entrepreneurship: Toward a Theory of Bureaucratic Political Power. The Organizational Lives of Hyman Rickover, J. Edgar Hoover, and Robert Moses (Bloomington: Indiana University Press, 1980), 58–63.

Early contributions to the AEC GOCO culture came from David E. Lilienthal, whose *Democracy on the March* (New York: Harper, 1944) reflects his concern to avoid a Washington-based bureaucracy. For a brief discussion of Lilienthal's distaste for centralized planning and his emphasis on democratic planning while serving as director of the Tennessee Valley Authority, see William Bruce Wheeler and Michael J. McDonald, *TVA and the Tellico Dam, 1936–1979: A Bureaucratic Crisis in Post-Industrial America* (Knoxville: University of Tennessee Press, 1986).

The local views of the risk at Hanford and the ways in which the nuclear culture there led to denial are spelled out in Michael D'Antonio, *Atomic Harvest: Hanford and the Lethal Toll of America's Nuclear Arsenal* (New York: Crown, 1993), and in a somewhat less sophisticated work by Michele Stenehejm Gerber, *On the Home Front: The Cold War Legacy of the Hanford Nuclear Site* (Lincoln: University of Nebraska Press, 1992). Another critical treatment of the nuclear engineers at Hanford is Paul Rogat Loeb, *Nuclear Culture: Living and Working in the World's Largest Atomic Complex* (Philadelphia: New Society Publishers, 1986).

The literature of international affairs which provides the background to the evolution of the weapons complex is vast. Several leading works which help place the nuclear weapon in the context of the cold war are Gregg Herken, *The Winning Weapon: The Atomic Bomb in the Cold War, 1945–1950* (Princeton: Princeton University Press, 1981); Martin Sherwin, *A World Destroyed: The Atomic Bomb and the Grand Alliance* (New York: Knopf, 1975); Philip L. Cantelon et al., eds., *The American Atom: A Documentary History of Nuclear Policies from the Discovery of fission to the Present*, 2nd ed. (Philadelphia: University of Pennsylvania Press, 1991); and Michael S. Sherry, *Preparing for the Next War: American Plans for Postwar Defense, 1941–1945* (New Haven: Yale University Press, 1977). The consequences of a command economy in the United States are described in William H. McNeil, *The Pursuit of Power: Technology, Armed Forces, and Society* (Chicago: University of Chicago Press, 1982).

The parallel history of power reactors has developed a large literature. One of the most balanced presentations is found in George Mazuzan and Samuel J. Walker, *Controlling the Atom: The Beginnings of Nuclear Regulation, 1946–1962* (Berkeley and Los Angeles: University of California Press, 1984). A work that explores the relationship of experts to public opinion and is more critical of the power industry and the AEC is Brian Balogh, *Chain Reaction: Expert Debate and Public Participation in American Commercial Nuclear Power, 1945–1975* (New York: Columbia University Press, 1991). In *Science, Politics, and Controversy: Civilian Nuclear Power in the United States, 1946–1976* (Boulder, Colo.: Westview Press, 1979), Steven L. Del Sesto traces the politics growing out of nuclear safety and shows that the AEC had an inadequate system for handling new issues. Spencer Weart, *Nuclear Fear: A History of Images* (Cambridge: Harvard University Press, 1988), shows how nuclear power reactors became a surrogate or symbol for the weapons. There are a host of less scholarly works that were published during the anti–nuclear power movement of the 1970s.

On probabilistic risk analysis, there is an obscure but vital literature reflecting the attempt to calculate the mathematical likelihood of nuclear disaster; see F. R. Farmer, "Reactor Safety and Siting: A Proposed Risk Criterion," *Nuclear Safety* 8 (1967): 539–48; and Chauncey Starr, "Social Benefit versus Technological Risk," *Science* 165 (1969): 1232–38. A practitioner's handbook which also provides background is Ralph R. Fullwood and Robert E. Hall, *Probabilistic Risk Assessment in the Nuclear Power Industry: Fundamentals and Applications* (New York: Pergamon Press, 1988); a fuller treatment of the issue in production reactor design is [Rodney Carlisle], "Probabilistic Risk Assessment in New Production Reactors: Background and Issues to 1991" (Office of New Production Reactors, Washington, D.C., October, 1992).

Thomas Kuhn, *The Structure of Scientific Revolutions* (Chicago: University of Chicago Press, 1962; 2nd ed., 1969), developed the concept of paradigm shifts in science, showing how professional institutions tended to resist changes to new viewpoints. The shift to probabilistic methods was accelerated by the Three Mile Island incident, which probabilistic methods would have anticipated better than deterministic methods. That anomalous event served to precipitate what Kuhn would call a "crisis stage" and wider acceptance of the new procedures. Another model for the shift in thinking might be found in Edward Constant, *The Origins of the Turbojet Revolution* (Baltimore: Johns Hopkins University Press, 1980), who traces the technological revolution of the aircraft jet engine. The paradigm change in risk analysis from determinism to probabilism, however, required uncomfortable rethinking among a cadre of reactor operators and designers, and their resistance bears striking resemblance to the conservatism of adherents of older scientific theories as discussed by Kuhn. George Basalla, in *The Evolution of Technology* (New York: Cambridge University Press, 1988), provides some very thought-provoking concepts regarding the evolution of technology from prior analogous tools, devices, or natural artifacts.

INDEX

Acheson, Dean, 68
Adams, Brock, 169 Table 8
Advisory Committee on Reactor Safeguards (ACRS), 102–5, 126–27, 128, 139–44, 158–59. *See also* Reactor Safeguard Committee
Aerojet General, 168
Ahearne, John, 191
Allison, Samuel K., 17
American Institute of Electrical Engineering (AIEE), 88
Anderson, Herbert, 20, 21
Argonne National Laboratory: origins of, at University of Chicago, 11; as site for X-10, 28, 43–44; location for move of CP-1, 32–33, 236n. 63; materials testing reactors at, 72; HWR design at, 78, 80–81, 84; move of testing by, to Idaho, 167–68; Maryland branch of, as office service contractor, 183. *See also* University of Chicago
Arms control treaties, 166 Table 7
Atomic Energy Commission (AEC), 3, 4, 10; origins of, 47–53; administration of nuclear facilities, 55–60, 99, 104; faces need for tritium for thermonuclear weapon, 70–75, 93–94, 96; safety concerns of, 84, 132–35; responds to peaceful goals with power generation, 107, 110, 117, 119, 128, 220; dealing with public regarding reactor closures, 142, 147–50, 251n. 47
Atomic Industrial Forum, 89, 144
Atomic piles. *See* CP-1; Power reactors; Production reactors; X-10; Test and experimental reactors
Atomic weapons. *See* Nuclear weapons

Bacher, Robert F., 48
Battelle Memorial Institute, 158
Beck, R. W., 121
Bloch, Edward, 103
Bloch, G. W., 144
Bonneville Power Administration, 120–21
Borden, William, 94
Brown, Harold, 175, 187
Browning, Joe L., 207–8
Bulletin of Atomic Scientists, 45
Burns and Roe, 120, 216
Bush, George, 214
Bush, Vannevar, 14, 52

Calvert Cliffs (Md.), 170
Carpenter, W. S., 19, 53
Carter, Jimmy, 160, 175, 202
Central Intelligence Agency (CIA), 67
Chernobyl reactor accident, 188–90
Church, A. E., 80–81, 87
Church, Frank, 169 Table 8
Church, G. P., 24
Churchill, Winston, 46
Cisler, Walker, 88–89
Clinton Laboratories. *See* Oak Ridge National Laboratory
Cochran, Thomas, 177, 181
Cole, J. E., 83–84
Columbia University, 14–16, 36
Combustion Engineering, 168, 211, 254n.18

Compton, Arthur Holly: as director of University of Chicago Metallurgical Laboratory, 13–16, 18–23; opinions on X-10, 28, 30; mediates scientist-engineer disputes, 33, 42
Compton, Karl, 52
Conant, James, 14, 21–22, 49
Cooper, C. M., 18, 19
Corporate culture. *See* Nuclear weapons manufacturing complex, culture of. *See also* Du Pont; General Electric
Costner, Brian, 212
CP-1 (Chicago Pile No. 1): first self-sustaining chain-rection in, 8; building and startup of, 20–23, 232nn. 31–35; move of, to Argonne Forest, 32–33; as demonstration of graphite moderation, 34; place of, in ancestry of reactors, 44
Craig, Larry, 169 Table 8

Daniels, Farrington, 43
Dean, Gordon, 74, 76, 94–95, 146
Department of Energy (DOE), 167, 175–78, 182. *See also* Nuclear weapons manufacturing complex, culture of; Office of New Production Reactors
Derrick, Butler, 169, 180, 262n. 1
Deuterium-moderated reactors. *See* Production reactors, heavy-water-moderated
Doan, R. L., 17, 29
Douglas United Nuclear, 140, 156
Du Pont (I.E. du Pont de Nemours, Inc.), as WW II reactor design contractor, 11–13, 22–35, 60, 95, 107; and WW II reactor operation, 41, 43, 44; departs Hanford, 53–55; as contractor for Savannah River, 74–75; flexible design methods of, 77–91; safety culture of, 138–44, 266; and HWCTR, 146; literature regarding, 266

EBASCO, 211
EG&G, 179, 253n. 9
Eisenhower, Dwight D., "Atoms for Peace initiative," 105–7, 110–11, 116, 130, 132; reorganization of AEC under, 113
Evans, Daniel, 169 Table 8, 190
Evans, R. M., 80, 82

Farmer, F. R., 140
Fast-Fuel Test Reactor (FFTR), 159–60, 227 Table A2
Fermi, Enrico, 15 ill.; and CP-1, 2, 8, 13–23, 38; participation of, in design of full-scale production reactor, 31, 37, 107; testimony of, at P-9 committee, 36; participation on GAC, 49, 69
Fermi reactor, 246n. 5
Fields, K. E., 99
Frankel, Ernst, 139
Fuchs, Klaus, 72–73

General Advisory Committee (GAC), of the Atomic Energy Commission, 49–50, 56–63, 69–71, 164
General Atomics, 121, 168, 177, 254n. 18
General Electric, as contractor at Hanford, 53–65, 78, 84–85, 92–110, 156; as builder of N reactor, 116–29; designer of experimental and export reactors, 168, 246n. 4; safety philosophy of, 247n. 27
German nuclear program in WW II, 1, 9
Gilbert, F. Charles, 176
Glennan, T. Keith 89; and issuance of Glennan report, 176–86
Gooding, Michael, 181
Gorbachev, Mikhail, 214–15
Gorton, Slade, 169 Table 8
Government-owned contractor-operated (GOCO) facilities: origins of, 10–11; transfer of, from MED to AEC, 51; effect in creating company town at Richland, 156; monopoly production and monopsony consumption by, 161–62; Watkins's reforms of, 202–3; evolution of political clout of, 219. *See also* Nuclear weapons manufacturing complex, culture of
Graphite-moderated reactors. *See* Production reactors, at Hanford. *See also* CP-1; Chernobyl reactor accident; X-10
Greager, O. H., 103
Greenewalt, Crawford: initial contacts of, with MED, 19, 21–23; direction of production reactor design effort, 25–37, 45, 107; in charge of reactor operation,

WW II, 42, 54–55; as president of Du Pont during Savannah River construction, 74
Greninger, A. B., 100
Groves, Leslie R., 49 ill.; rapid procedures of 6, 9, 20, 23, 37; military leadership style of, 26, 45, 47; shaping nuclear weapons complex culture, 11–13, 50–53, 202; decisions of, regarding X-10, 28, 30; obtains P-9 committee review of heavy water reactor design, 34–36; role of, in engineer-scientist dispute at startup of B reactor, 37–39; position of, in GE takeover of Hanford, 54–57
Gunn, Ross, 44

Hahn, Otto, 14
Hall, A. E. S., 24
Hanford, Washington, 8, 10, 12, 155 map; choice of site, 20, 23–25, 29, 30; construction at, 36–38; early operations at, 38–42; local organizations at, 188. *See also* Production reactors, at Hanford
Hart, Gary, 177
Hatfield, Mark, 195–96
Heavy Water Components Test Reactor (HWCTR), 146
Heavy water reactors (HWRs). *See* Production reactors, heavy-water-moderated
Hekman, Peter M., 205
Hendrickson, Thomas, 216–17
Herrington, John, 187–93, 197, 202, 205, 220
High-temperature gas-cooled reactors (HTGRs). *See* Production reactors, high temperature gas-cooled
Hilberry, Norris, 17, 19
Hillenkoetter, Roscoe, 67
Hodel, Donald, 178–79, 182–86, 190
Holifield, Chet, 123, 247n.18
Hollings, Ernest, 169 Table 8
Hollingsworth, R. E., 150
Hosmer, Craig, 122

Idaho National Engineering Laboratory (INEL): origins and extent of, 167–68, 253n. 2; production reactor replacement office established at, 175; experience with experimental reactors at, 179; Senator McClure's advocacy of, as site for new production reactor, 182, 183, 185, 194
Indian Head Naval Ordnance Station (Md.), 11, 207
Iotti, Robert, 211

Jackson, Henry, 118–21, 150, 169 Table 8
Johnson, Lyndon, 148, 158, 161
Johnson, Wilfrid E., 116, 118, 119
Joint Committee on Atomic Energy (JCAE): as forum for nuclear issues, 50–51, 53, 58, 164, 168; dealing with GE as contractor, 58, 65; concern with funding, 61; decision of, to build thermonuclear weapon in response to Soviet advances, 67–72; selection of Savannah River site by, 74–75; response of, to Korean War, 93–96; N reactor decisions by, 111, 117, 122, 128; and alternate uses of HWRs, 144, 146; and closure of reactors, 150, 156

Kempthorne, Dick, 169 Table 8
Kennedy, John F., 113
Khrushchev, Nikita, 148
Knolls Atomic Laboratory, 54, 78, 81, 84, 109–10

Lawrence, Ernest O., 13, 69
Lawrence Livermore Laboratory, 1
Le Baron, Robert, 94
Lee, Glenn, 137, 150, 153–54, 162, 219
Lehman, John F., 200
Leventhal, Paul, 193
Leverett, Miles, 17
Lilienthal, David E., 48, 49 ill., 51–52, 56, 68–70
Loper, Herbert, 73
Los Alamos National Laboratory, 11, 13; need for reactor products at, 29, 42, 43; in NPR site dispute, 182

Magnuson, Warren, 150, 169 Table 8
Manhattan Engineer District (MED), 3,

Manhattan Engineer District *(cont'd.)*
 9–11, 17–18, 48, 51–52. *See also* Groves, Leslie R.
Mark, J. Carson, 193
Marshall, James C., 17
Matthais, Franklin T., 24, 37, 233n. 44
Mattingly, Mack, 120
May, Andrew Jackson, 48
McCarthy, Joseph, 132
McClure, James A., 168, 169 Table 8, 182–85, 194–96, 262n. 1
McCormack, Mike, 169
McCullough, C. Rogers, 102, 144
McFarlane, Robert, 187
McMahon, Brien, 48, 50, 69, 70, 73, 93–94
McNamara, Robert, 198
Meitner, Lise, 14
Met Lab. *See* University of Chicago
Military Liaison Committee (MLC) of the Atomic Energy Commission, 49–50, 57, 59, 93, 94, 164
Milliken, Eugene, 67
Monetta, Dominic J., 206–17, 220, 245n. 1
Monsanto Corporation, 30, 53
Moore, Thomas, 17, 18, 43
Morrison, Sid, 169, 190, 196
Murray, Patty, 169 Table 8
Mutual assured destruction (MAD), 165

Nader, Ralph, 171–72, 203
Nelson, Curtis, 85
New production reactors (NPR), dispute over location of, 174–87, 253n. 8. *See also* Office of New Production Reactors (ONPR)
New York Shipbuilding Company (Camden, N.J.), 87–88, 89 ill.
Nichols, Kenneth D., 43, 73
Nuclear culture. *See* Nuclear weapons manufacturing complex, culture of
Nuclear espionage, 1, 48, 72–73
Nuclear power. *See* Power reactors
Nuclear reactors. *See* CP-1; Power reactors; Production reactors; Test and experimental reactors; X-10
Nuclear Regulatory Commission (NRC), 167
Nuclear weapons, 2, 3, 26, 42, tests of, 96; proliferation of, to new states, 222–23. *See also* Thermonuclear weapons
Nuclear weapons manufacturing complex, culture of: shaped by Groves's decisions, 3–4, 9–14; Watkins's attempts to modify, 7, 196–206; Lilienthal's influence on, 51–52; isolation of, from nuclear power culture, 90–91; tradition of secrecy in, 131–33; modernization report, 192–93; evolution of, summarized, 219–20; scientists absorbing craft culture in, 230n. 10; accusations of complacency in, 259n. 20; theoretical framework for discussing, 265–67
Nunn, Sam, 216

Oak Ridge National Laboratory: X-10 reactor at, 27–30, 39, 44, 53; participation of, in HWR design, 78, 84
Office of New Production Reactors (ONPR), 205–17, 220–22; DP 13 and DP 132 as predecessors to, 183
Office of Scientific Research and Development (OSRD), 14, 16, 17, 18, 52, 202
Oppenheimer, J. Robert, 4, 13, 42–43, 50, 69

P-9 Committee (at Chicago Met lab, to review conceptual design), 32–37
Pastore, John, 156
Patterson, Robert P., 47, 53
Peters, Thomas J. 196
Phillips Petroleum, 168
Pike, Sumner T., 48, 70
Pilot plant. *See* X-10
Plutonium: for WW II uses, 8, 26, 29, 42, 55; postwar production of, 56, 57, 62, 65; balancing production of, with tritium, 70–73, 86, 94; eventual surplus of, 76, 146–48, 158–61; production of, in N reactor, 113, 116–19, 128; increased requirement for, 175; conversion of power reactor to produce, 190; as reactor fuel, 159–60; monopsony consumption of, 160
Polonium, 42–43, 55–56, 157, 238n. 21
Power reactors: as descended from production reactors, 2; safety controversies sur-

rounding, 5, 90–91, 135–36, 171; AEC promotion of, 111–13, 114–15 Table 4; total power of, compared to production reactors, 153 Fig. 2, 248n. 40, 251n. 48; WNP reactors built by WPPSS at Hanford, 160, 173; literature regarding, 267. *See also* Chernobyl reactor accident; Fermi reactor; Shippingport reactor; Three Mile Island reactor accident; Windscale reactor accident

Price, Byron, 120

Price, Melvin, 184–86

Price-Anderson bill, 170–71

Probabilistic risk assessment (PRA). *See* Production reactors, probabilistic risk assessment, in design of

Production reactors

—conceptual designs of: University of Chicago groups work on, 16–17; X-10, 26–30; Du Pont decisions on, for first production reactor, 31–32; P-9 committee debates over, 32–37; alternate, during WW II, 43–45; KE and KW as following original, 99; heavy-water, by Walter Zinn, 129–30; technopolitical debate over, 174, 176–80; as handled by ONPR, 210–15

—convertible to power, in Europe, 115, 121

—effluents from, 6, 30–32, 64, 83, 97, 129, 191, 248n. 4

—families of, 2, 5, 43–44, 92, 114–15, 129, 152, 225–28, 226 Table A.1

—at Hanford: B, D, and F reactors, 26, 32, 37–44, 40 ill., 47, 55–62, 65, 76, 92, 148; scrams of, 42 table 1; dates of completion of, 78 Table 2; DR reactor, 59–61, 68, 72–73, 92, 110, 148; H reactor, 59–61, 72–73, 92, 94, 99, 110; C reactor, 95, 96, 151; KE and KW ("Jumbo") reactors, 96–101, 107, 110, 151; power levels, 104 Table 3, 225–27; N ("convertible") reactor, 107–30, 112 ill., 151, 173, 220, 246n. 9; N charging face, 125 ill.; N control room, 127 ill.; closures of, 152 Table 6; N compared with Soviet RBMK type, 188–90; naming of, for letters, 235n. 45

—heavy-water-moderated (HWRs), 33–36, 44, 72, 83 ill., 89 ill.; 145, 210. *See also* P-9 Committee; Production reactors, at Savannah River

—high-temperature gas-cooled (HTGRs), 43, 176–179, 182, 193–94, 210, 254n. 18, 260n. 38. *See also* General Atomics

—planning for, affected by world conditions: Cold War needs, 47, 56–60, 98 Fig. 1; Soviet nuclear weapons, 65–77, 92–99; Cold War thaw, 105–8; during MAD, 165–66; Afghanistan, 175; as bluff, 187, 262n. 2; end of Cold War, 219–21

—power levels of: on original WW II reactors, 40–42; postwar upgrades of, 60–64; at Savannah River, 81–82, 86; new scale of, for new round at Hanford, 92, 95; of KE and KW, 97; safety concerns during upgrades of, 102–5; Hanford upgrades of, by 1958, 104 Table 3; upgrades of all, by 1963, 147 Table 5

—probabilistic risk assessment, in design of, 6, 124–26, 138–41, 170–71, 268

—risk of, 58, 63, 90, 100–105, 123–27, 129, 132–43

—at Savannah River, 73–91, 82 ill., 146–51, 173, 227–28; K reactor, 191, 206, 213–14; L reactor restart controversy, 180–82

—slug failures in, 31–32, 52, 64, 65, 83, 100–103, 116, 239n. 41

—total power of, compared to power reactors, 153 Fig. 2, 248n. 40

—xenon-poisoning in, 38–41, 51

Prout, G. R., 65

Quarles, Donald, 88–89

Rabi, Isidor, 49, 69

Raborn, William F. 198, 207

Rasmussen, Norman, 170–71

Reactor families. *See* Production reactors, families of

Reactors, production. *See* Production reactors

Reactor Safeguard Committee (RSC), 63–64, 100–102, 137. *See also* Advisory Committee on Reactor Safeguards (ACRS)

Reagan, Ronald, 168, 176, 183, 202
Resource Alternatives, Inc., 217
Richland, Wash., 151, 154, 155 map, 156, 162. *See also* Production reactors, at Hanford
Rickover, Hyman, 109, 111, 198–201, 208–9, 258n. 13; literature regarding, 266–67
Roddis, Louis, 189
Russell, Donald, 149
Russia. *See* Union of Soviet Socialist Republics

Salgado, Joseph, 190, 258n. 16
Savannah River, S. C., 74–76, 78 Table 2. *See also* Production reactors, at Savannah River
Schlemmer, Frederick, 60, 68
Schneider, Keith, 192
Scrams: origin of term, 22; in early reactor operations, 42 Table 1
Seaborg, Glenn, 17, 50, 150, 156, 158
Shaw, Lewis, 197
Shaw, Milton, 246n. 3
Shippingport reactor, 109, 111
Shugg, Carleton, 60, 68
Sierra Club, 180–81
Smith, Levering, 198, 207
Smyth, Henry D., 33, 70, 73
Smyth Report, 2
Somervell, Brehon, 18
Space Nuclear Auxiliary Power (SNAP), 144–45
Spedding, Frank, 17
Spratt, John M., 212
Stalin, Joseph, 105
Steele, Karen Dorn, 188
Stone and Webster, 17, 120
Strauss, Lewis L. 48, 68, 69
Suits, C. G., 94
Symms, Steven, 169 Table 8
Szilard, Leo: anticipates weapon, 12; conceives reactors with Fermi, 14, 15; at Chicago, 17, 23, 33, 45, 237n. 65; testifies at P-9 Committee, 35

Technopolitics. *See* Production reactors, conceptual designs of

Teller, Edward, 63–64
Test and experimental reactors, 227 Table A.2. *See also* Idaho National Engineering Laboratory
Thermonuclear weapons, 68–73. *See also* Production reactors, planning for, as affected by world conditions
Three Mile Island reactor accident, 172, 188
Thurmond, Strom: as president pro tem of Senate, 168; as influential in NPR siting controversy, 169 Table 8, 179, 184, 185, 187, 194, 262n. 1; concern of, with L reactor restart, 181
Tolman, R. C., 37
Tower, John, 185, 186
Tri-City Nuclear Industrial Council, 154, 156, 157, 190
Tritium: decay of, as automatic disarmament, 5, 6, 193; balancing production of, with plutonium, 70–73, 76, 86, 91, 94; production of, in N reactor, 128; continuing need for, 147, 158, 190; monopsony consumption of, 160–61, 190
Truman, Harry S.: sets Cold War policy of nuclear armament, 46–48; concern with inadequate nuclear stockpile, 56; decision of, to proceed with thermonuclear weapon, 68, 70; asks Du Pont to participate in new construction, 74; reacts to North Korean invasion of South Korea, 93, 95–96
Tuck, John, 217

Union of Soviet Socialist Republics (USSR): American public perception of threat from, 3–4; Berlin blockade by, 60; U.S. reaction to first nuclear test by, 67–70, 92–99; RBMK reactors in, 115; dual purpose reactors in, 121; and MAD, 165–66; Carter's reaction to Afghanistan invasion by, 175; Chernobyl accident in, 188–189; decline of Cold War tensions of, with U.S., 214–15, 219
University of California, 3, 11. *See also* Los Alamos National Laboratory
University of Chicago, 3, 11–18, 23, 28–39,

45. *See also* Argonne National Laboratory; CP-1
Urey, Harold, 34–35

Valente, Capt. (diarist at Hanford), 39–40
Wade, Troy, 180
Washington Nuclear Power (WNP) reactors. *See* Washington Public Power Supply System
Washington Public Power Supply System (WPPSS), 123, 156, 159, 160, 189
Waterman, Robert H., 196
Watkins, James, 195–207, 212–14, 216, 220
Waymack, William W., 48
Weaver, James, 177, 189
Weinberger, Casper, 187
Westinghouse Corporation: in early power reactor development, 111, 112, 121, 176–77; at Hanford, 156, 190; experimental reactors of, 168; at Savannah River, 207
Wheeler, John, 17, 28–29, 79, 235n. 50, 236n. 54

Whitaker, Martin D., 29, 30, 35
Wigner, Eugene: participation of, in early reactor design, 16, 17, 18, 21, 23, 29; advocate of HWR on P-9 committee, 33–37; discovery of graphite expansion effect by, 58; participation of, in planning Savannah River reactors, 78. *See also* Windscale reactor accident
Williams, Roger, 28, 29, 33–35, 37
Williams, Walter J., 56–57, 60
Wilson, Caroll, 52, 60, 61
Wilson, Charles, 54
Windscale reactor accident, 102–3
Winne, Harry A., 58
Worthington, Hood, 49, 85

X-10 (reactor at Oak Ridge), 26–31, 43–44

Yeltsin, Boris, 215

Zinn, Walter, 20–21, 39–40, 78, 80–81, 129

Library of Congress Cataloging-in-Publication Data

Carlisle, Rodney P.
    Supplying the nuclear arsenal : American production reactors.
1942–1992 / Rodney P. Carlisle, with Joan M. Zenzen.
      p. cm.
    Includes bibliographical references and index.
    ISBN 0-8018-5207-2 (alk. paper)
    1. Nuclear reactors—United States—History. 2. Nuclear Weapons—
Equipment and supplies—Government policy—United States.
3. Nuclear fuels—United States—Breeding—History. I. Zenzen,
Joan M.
TK9202.023 1996
355.4'3'00973—dc20   95-44410

www.ingramcontent.com/pod-product-compliance
Lightning Source LLC
Chambersburg PA
CBHW021119300426
44113CB00006B/218